Text, Translation, Computational Processing

Edited by
Annely Rothkegel and John Laffling

Volume 10

Communication on and via Technology

Edited by
Annely Rothkegel and Sonja Ruda

DE GRUYTER
MOUTON

ISBN 978-3-11-026025-0
e-ISBN 978-3-11-026027-4
ISSN 1861-4272

Library of Congress Cataloging-in-Publication Data
A CIP catalog record for this book has been applied for at the Library of Congress

Bibliographic information published by the Deutsche Nationalbibliothek
The Deutsche Nationalbibliothek lists this publication in the Deutsche Nationalbibliografie; detailed bibliographic data are available in the Internet at http://dnb.dnb.de

Typesetting: Meta Systems GmbH, Wustermark
Printing: Hubert & Co. GmbH & Co. KG, Göttingen
∞ Printed on acid-free paper
Printed in Germany

www.degruyter.com

Contents

Annely Rothkegel and Sonja Ruda

Introduction

Technology and communication are closely related to each other. This anthology focuses on the role of communication. Without it, in the exchange of knowledge, there would be no technical innovations. The use of technology, supported by documentation, would not be successful without communication. Similarly technology supports communication, which – independent of time and space – develops into new forms. The globalized trade brings about questions that stress the characteristics of languages and cultures at the intersection of technology and communication. Technical artifacts play an important, often central role in almost all areas of life. In both everyday life and the professions we live in contact with technology in form of its products. They determine the course of our actions, and not seldom even our objectives. Simultaneously communication communities establish which, in turn, generate communicative products and use or work with them. The scientific study of the phenomena resulting in this tension opens up horizons that cut across conventional disciplines and combine them.

The dealing with technological knowledge in documented communication can be traced back to the distant past. Thus present technical documentations combine traditions of descriptions with legal and normative requirements on the one hand and with standardization tendencies based on the software tools employed for the documentation work on the other hand. The professionalization of the documentation work has produced two threads of interest. These are the practice (in and for companies) as well as research in the disciplines which are thematically involved. Further included are Linguistics, Translation Science, Media Science, Intercultural Communication, Cognitive Psychology, Sociology, Computer Science, Mathematics, Technological Philosophy. The different perspectives on the reference range "industrialized real world" bring into bear characteristics of communication that are indispensable for dealing with technical artifacts. They are clustered in three lines: the transfer of technical knowledge through communication (Chapter 1), the role of sign systems like language and visual images for the transfer of information (Chapter 2) and finally the relations between technology, culture and computerized communication (Chapter 3).

Chapter 1 addresses communication problems related to the transfer of knowledge. The article by **Annely Rothkegel** examines the tension of communication on technology between theory and practice. A variety of meeting points are discussed from a text-linguistic perspective: knowledge in the mix of expert and common knowledge, acting with and through technology, texts as communicative means of transferring knowledge as well as the cultural character of using and documenting technology.

In a cognitive-psychological approach, **Frank Ganier** first outlines different models for the processing of procedural instructions which are understood as a component of a user document equipment interaction. His own research yields the interactive model which – unlike the linear concepts of reading, understanding and the subsequent performing of action steps – consists of a complex mental model of previous knowledge, action plan and the monitoring of error corrections. Apart from theory development the model also aims at improving training for the practice of writing and reading technical documents.

The question of what kind of knowledge is regarded as necessary and preferable in instruction texts or the way of determining this knowledge is explored by **Sonja Ruda** from a linguistic-cognitive perspective. She outlines a method for ascertaining knowledge and problems which can be employed to systematically carry out expert and user interviews and integrate them in terms of the ascertained data. The resulting transcriptions and documentations constitute a reliable basis for the right questions and answers which, in turn, form the foundation for creating user-oriented instruction documentations.

Knowledge production, closely related to text production, is the topic of the sociolinguistic approach of **Karl-Heinz Pogner**. He applies the comparison of the concepts of discourse and practice communities to three case studies in the context of Danish enterprises and a university institute. This results in the realization that, depending on their type, communities function differently as contexts for the text and knowledge production. Consequences arising for the practice at the workplace on the one hand and for research on the other hand are taken into consideration.

Chapter 2 includes four articles focusing on the conditions of the linguistic or visual construction of meaning for the presentation of knowledge and the transfer of information. The first two studies deal with the monolingualism or multilingualism of documents and their creation. From a linguistic and translation-theoretical point of view, **Klaus Schubert** discusses the close relations between technical writing and technical translation elaborating their common characteristics and employing the term “technical communication” as a joint umbrella term. This view, already established in practice, is conceptually and theoretically further differentiated here. Based on the drafted model with the four key points “technical content”, “linguistic form”, “technical media” and “working processes” and their interdependencies, both fields are systematically brought together. This is verified by the detailed analysis of selected problem areas such as wording, software localization, subtitling or the reusability of linguistic units.

The micro level of language usage is emphasized in **Colette Cortès'** article. She examines the conditions under which adjectives unfold their class-building potential thus showing the orders of the world knowledge presented in the text. The research corpus relates to a variety of articles out of a catalog for everyday products which also include technical products. The analyses, carried out using classical linguistic methods, illustrate that – similar to instruction manuals – object types rather than single objects are described. On the other hand adjectives are also used for evaluating-advertising functions which relate to specific speech situations. This results in a mix of technical and everyday language typical for the expert – nonexpert communication.

Clemens Schwender addresses the characteristics of visual images as sign systems by tracing the illustration techniques of visual instructions in their historical development. Accordingly, all presently used principles of text visualization were introduced as early as in the time of Leonardo da Vinci and Georgius Agricola. This includes reference systems in the image, axonometric projections or techniques for the depiction of time, movement, direction or intensity. The increasingly used pictograms, prototypes for the visual depiction of actions, are likewise regarded as "old" inventions. "New" elements, on the other hand, can be found in the field of visualization technologies and terminology.

In Chapter 3, technology is related to communication and culture(s). Thereby the hitherto prevailing view on technology as an autonomous field unravels. This becomes evident with a view to the communication technology and the computer as well as the corresponding software as technology involved in all areas of life which, in turn, is imparted communicatively, e.g. through software documentation. Influenced by a mathematical background, **Vasco Alexander Schmidt** argues the case for the classification of dealing with computers as the fourth basic culture technique apart from reading, writing and calculating. Thereby it needs to be considered that software as a technical product requires medial communication itself in order to become visible and thus manageable. This allows the connection of a specific domain (computer/software) with another specific domain (e.g. administration, scientific disciplines, communication etc.) in the first place. The task to mediate the thus resulting complexity of the software application on the basis of concurrent functions of the computer as a tool and medium and, moreover, as an instrument of gaining knowledge poses special requirements on the communicative competence. This, in turn, is likewise demanded when it comes to the configuration of user interfaces for the human-computer interaction or the reception of information imparted by them.

The application of culturally related dimensions to technology and technical documentation is explored in the article by **Marc Hermeking**. Based on examples and an approach of intercultural communication, he disproves the assumption that standardized technical products create a homogenous "one-world" culture. Emphasis is laid on the computerized communication and, in particular, the interfaces of human-computer interaction. Ways of using the internet or the www are examined in view of the culturally determined dimensions of time, space and interpersonal interaction (high-context cultures vs. low-context cultures).

At the close of the culturally-oriented chapter and as an outlook of this anthology as a whole, **Gerhard Banse** and **Andreas Metzner-Szigeth** introduce a research network which is concerned with a trans-European survey into the influence of the internet as a socio-technical system on the changes of socio-cultural practices in the different regions. New media and their relation to the cultural diversity in Europe constitute the survey terrain which reveals different forms of community building. Four subject levels are brought into focus: privacy and the public sphere (socio-political dimension), identity and the community (socio-cultural dimension), knowledge and the economy (socio economic dimension), and cut across them security/insecurity and trust. Within this frame with desired disciplinary diversity, research aims at developing a research matrix whose application results facilitate the deduction of guidelines for political acting in support of promoting European integration.

The assembled articles draw a connection from the micro to the macro level of documentation as well as from individual to social dimensions of communication on technology on the whole. The joint consideration of technology as a socio-communicative phenomenon opens up new research questions with an interdisciplinary horizon.

I. Technological knowledge in communication

Annely Rothkegel

Communication on technology: Meeting point of theory and practice

1 Approach

1.1 Technology and communication

Technology evokes communication. Or is it the other way around? No matter in which direction we look at the dynamic, both are closely linked (Bammé 2008). Going briefly back to the technology descriptions in the craftsmen's books of medieval times or later to the thousands of drawings of technical equipments out of Leonardo da Vinci's (1452–1519) workshops or maybe to the likewise artistically crafted woodcuts drafted by Georgius Agricola (posthumous 1556, 2003 reprint of VDI-edition of 1928) together with explanatory texts describing the activities in the then prospering ore mining, communication in these examples primarily stands for the transfer of competences for the reproduction of the technical innovations or for the steps necessary to melt and forge ore. Schwender (2003) sketches such historical threads of technology description which on the one hand can be further traced back to 2000 BC (Schwender 2003: 35) and on the other hand reach in continuous lines to the present day (also see Schwender's article in this volume: Schwender 2012).

Today it is legally required product documentations that, for reasons of liability, are considered as parts of a technical product and thus arouse increasing attention in practice and training. User orientation in the selection of knowledge and comprehensibility for the users as readers are becoming quality requirements of communication. In addition, public debates are relevant for the acceptance of technical innovations. Thereby it is the debate itself which determines the direction of the technology development. One example is the change of thinking in the 1970s when, in consideration of a non-acceptable number of accidents, the automotive industry was looking for technical solutions that provide more safety. Terms like active and passive safety have become central concepts of technology development (Stieniczka 2006) which are today reflected in the sensor- and computer-controlled assistance systems. The understanding of the changed roles of the driver as actor and/or co-driver and the technical system as co-driver and/or actor has to be imparted communicatively.

The safety issue clearly illustrates that and how technical problem solutions evoke new problems in turn. This does not only concern the prevention of technically generated risks (concept "safety") but also the measures directed against intended attacks during and caused by the use of technology (concept "security"). The evolving situations necessitate a successful communication between actors, participants and those concerned (keyword "body scanner"). In this regard it is particularly the mix of specialist knowledge from different fields and situational knowledge including all possible contexts that makes communication on technology interesting.

1.2 Actors of communication

Talking or writing about technology normally requires professionalism. On the one hand it is about technical issues that constitute the reference range in the particular communication situation and are connected to specialist technical knowledge. On the other hand it involves communication skills which exceed specific technical knowledge and are part of the professional practices of media-oriented working environments. Insofar as communication on technology is equivalent to communication on technical products and their use, further contexts come into play. The interaction between human and machine, the humans' handling of artifacts come to the fore.

The field of technology and the use of its artifacts affect (almost) all areas of life. In both everyday life and profession we live in contact with technology in the form of its products. They mark the course of our actions and sometimes even determine our objectives. Simultaneously communication communities establish where actors operate in different roles: decision makers, development engineers, seller, user and consumer, technical authors. Technology development, distribution, use and disposal are inconceivable without communication. For safety reasons legally required, advertised for marketing purposes, as part of curricula or instruments of knowledge transfer – communicative products (texts, images and videos, multimedia products) are gaining increasing importance. Their specific characterisation is, however, not merely based on their technical relatedness but rather on the fact that both threads which accrued from practice – centuries-old tradition as well as modern standards of technology documentation – are increasingly subject to a scientification of professional operations.

1.3 Science and practice

Reversely, in particular Applied Linguistics or Communication Science has increasingly (since the pragmatic turn) attended to questions which refer to specific communication patterns and communication forms in real situations (Knapp et al. 2011). This, in turn, is not uninteresting for practice insofar as one expects knowledge about more effective strategies in dealing with miscommunication or with communication problems in general (Fiehler 1998). However, there is a catch: theory and practice generally pursue different goals. Their questions are not the same. Pessimists therefore believe that theorists and practicians actually have nothing to communicate about. The present optimistic article employs a number of central concepts (communication, knowledge, action, communication form, cultural aspects) to examine whether and where there are yet any contact points in the diverse.

What is different? Science asks: What is the case and why is it the way it is? How can this be observed, examined, proved? For the former, theories are developed, for the latter, models and methods are worked out. New insights (e.g., through new instruments) lead to new theories, models and methods. Generalisation at the conceptual level is an important principle. It involves universal statements which are independent from the individual case but ideally applicable on all individual cases. Practice asks: What is good or better for which purpose? What advantages or disadvantages does A have against B? Here it is about experience in individual cases which can be used to derive beneficial problem solutions ("best practice"). They ensure reliability and can be formulated for repeated applications as "acknowledged standards of technology". They are expanded into norms and benchmarks, determine procedures and quality and prescriptively promote standardisations (Hackos and Stevens 1997; Brandt 2005; Herzog and Mühlbauer 2007; DIN Taschenbuch 2008; Frei 2008; Muthig 2008).

2 Communication and communication models

2.1 Two perspectives

The concept of communication is so ambiguous that it would actually be useless if it did not have a simple core: sender and receiver are connected with each other by means of a content which is sent or received. Being a technical term, it is marked by theories or models of the discipline where it is used. A distinction is made between a technically understood communication concept and a linguistically determined communication concept including the characteristics of language as a sign system.

Let us first consider the technical model of communication. As far as to the present discussion it has been based on the mathematically funded model by Shannon and Weaver ([1949] 1995). Apart from sender and receiver, the model also differentiates the components of the channel (carrier medium) and potential disturbances (noise). The contents are specified as signals. This model, often supplemented by the aspects of feedback and loops, continues to be employed for descriptions of information flows in business management. It tracks the movement of information (information flows) from point to point except for the contents as such and the changes in the actors themselves, i.e., everything occurring with regard to sender and receiver.

The content itself as well as its handling and the mutual impact the actors exert on each other are, however, of interest in scientific models of interpersonal communication (overview in Krallmann and Ziemann 2001; Schmidt and Zurstiege 2007). Also this field has been characterized by a model dating back to the first half of the 20th century, Karl Bühler's communication model (1934 or 1982 or 1999). It is based on a language theory which puts the linguistic usage into the centre of attention. Base unit is the sign, a relation between form, content and the sign user. The form can include units such as words, sentences, text components or whole texts. The contents correspond to the represented or, in other words, the reference range. Here, the symbolic character of language becomes apparent. Sender and receiver are involved in the way that each utterance additionally transmits something about the sender himself (symptom character), e.g., his attitude towards the content and the receiver. Moreover the utterance implies an impact on the receiver (appeal character). This mutual reference of sender and receiver in relation to the represented (content) becomes itself object of modelling in the speech act theory (particularly in the approach of illocutionary classes by Searle (1969), which shall be examined separately in section 4). The social relation between the actors participating in the communication situation is of relevance in extended models (e.g., Schulz von Thun 2004) which go beyond Bühler and include psychosocial parameters (e.g., partnership, dominance, individuality, group membership, identity).

An important distinction in both approaches involves the view on both the content and the characteristics of sender and receiver. From a communication-technological perspective, a certain consistency of the components can be assumed. The transfer of the signal may be disturbed, i.e., it may not reach the receiver in the sent form, though without a disturbance (noise) of the channel or from outside, a "correct" reproduction can be relied on. From the point of view of communication science on the other hand, contents are concerned with meaning or attribution of meaning, apart from the fact that disturbances in the acoustic and visual range may also occur here, which are, however, not being referred to here (Lenke, Lutz and Sprenger 1995; Faßler 1997; Brosius

2005; Misoch 2006; Burkart 2007). We are referring to the fact that it must be assumed that sender and receiver generally act with different constructions of meaning despite using the same wording. This is on the one hand founded on the symbolic character of language and, on the other hand, on our cognitive conditions of reception and comprehension. Each of us develops specific ideas (models) about the "world" which are related to our experience and learning processes. In contrast to the communication-technological approach, the linguistic-communicative approach acts on the assumption that the contents uttered on the part of the sender are not identical with contents viewed by the receiver. From a constructivist point of view communication is not regarded as a transfer or exchange but rather a transformation (Gerstenmaier and Mandl 1995). This fact casts light on how we can imagine comprehension and intelligibility as a possibility of re-construction against the background of the respective individual knowledge. From this perspective it is the sender (e.g., the author) who has to consider such possibilities and avoid unwanted options. As far as the relation between sender and receiver is concerned, special emphasis is accordingly placed on impact and change dependent on communication.

2.2 Topics of theory and practice

Our understanding of communication determines on the one hand the selection and the handling of approaches in research and theory formation and on the other hand practical tasks and problem solutions. In this comparison we will look at the following four dimensions within which theory meets practice:

(a) Knowledge: Communication on technology is characterized by the mix of expert knowledge from various fields with situational knowledge from the application areas of everyday life and the professions. The question that poses itself is about the strategies to capture this knowledge with reference to the most different contexts. Of central importance is the way the applied terms are handled and how they are understood in the different contexts. One objective of knowledge communication involves conceptual transparency as a prerequisite for the creation of comprehension.

(b) Actions: The communicative enabling of actions first of all aims at the practice of human-technology interaction. From here, the question arises which concepts for "acting with technology" can be regarded as relevant. Communication on technology involves two action levels: practical actions of using a product and speech acts as communicative actions. The latter assign the former to a modus operandi, e.g., in the instruction on what needs to be done to attain the targeted positive result or the warning about what is (not) to be done to avoid dangers and damages.

(c) Texts as a mediation form: The symbolic transfer of knowledge and communicative actions requires organization and form. Texts, spoken or written, printed or online, are regarded to be conventional forms of organizing communication. Research is interested in its structures, characteristics and processes when being dealt with by writers and readers as users. In practice, aspects of tool usage (software) and management of data and formats are given priority. Increasing standardization promotes this tendency.
(d) Cultural aspects: Knowledge, actions and the forms of organization which impart them are determined by culture. If we characterize "culture" as "the self-evident", "cultures" become a comparison of what is noticed as being "different". The transfer of technical knowledge and technical products around the world implicates the problem of identifying the self-evident or striking in each of the different discourse and practice communities.

Individual discussions examine the meeting of theory and practice in a problem-oriented way. The following questions are of central interest: Do the research questions originate from theory or practice? What scientific instruments are relevant for practice? What role do analysis and production play? Can description and prescription be divided?

3 Knowledge: structuring and description

3.1 Systems in research and practice

Communication on technology is mainly about human-object relations with and in their different contexts. It involves numerous scientific disciplines, each dealing with differently formulated principles, structures, processes or procedures and methods. Each on its own is conceptually organized in a particular way. Knowledge appears in more of less stable systems of concepts which form taxonomies of a specific field and are subject of continuous discussion. Although findings and approaches are borrowed from natural sciences (and mathematics), thinking is directed towards artefacts, i.e., objects that were or are to be intentionally created. From a practical point of view, these artefacts become technical products and eventually goods which are used, consumed and disposed of or recycled. It is referred to product life cycles (conception, construction, production, distribution, consumption, disposal) bringing together objects and humans in very different contexts (Mansour 2006). The objectives of the knowledge communication taking place here are in both theory and practice directed towards order and understanding. In science this is imperative for the generation of knowledge. Both are important in practice for the transformation of knowledge into practical actions.

Where are the interfaces? Let us look at the forms of organization which are employed to describe knowledge. An established and common form is hierarchical classification systems which records relations between hypernyms and hyponyms. Such arrangement systems describe subject areas according to different criteria (taxonomies). Knowledge is structured through definitions, i.e., by specifications and grouping on the basis of distinctive characteristics. Just as these taxonomies, hierarchies on the basis of semantic part-whole relations (meronomies) are likewise helpful for the demarcation of relevant reference ranges in a respective subject. Moreover they are very suitable as practical classification systems in operational applications (stock keeping, administration of bills of materials, catalogue maintenance). As ontologies (classification systems of "that which is being") they structure worlds or certain ideas about them, which are expressed in the classifications or segmentations into partial units (King and Reinold 2008).

3.2 Domains and terminology

The attempts to represent a coherent system of concepts for the vocabulary of a language in a classification of that kind, as done in previous onomasiological approaches, were criticized as subjective and generally abandoned in the end. The construction of such structures is not trivial. Particular importance is attached to the selection of the class-distinguishing features and the determination of the kind of units. Furthermore, problems of cross-classification and incompleteness (remaining classes) must be taken into account as well. The restriction to a delimited domain and the inclusion of different relations, typically referred to as thesaurus (Wendt 1997), can facilitate a structured survey of a subject area within defined limits. This is the claim raised in LSP-research (Language for Special Purposes, overview in Roelcke 2005 and Schubert 2007) and particularly in the science of terminology (Hoffmann, Kalverkämper and Wiegand 1998/99; Arntz, Picht and Mayer; 2004; Picht 2006). Problems of terminology expanded into practice first in the field of translating and later also with regard to single languages in the area of technical documentation (Hennig and Tjarks-Sobhani 2008).

A discipline or a branche are (ideally) systematically structured in their subject area and the assigned terminology. That is what makes up a domain and distinguishes it from other specific fields. The domains are differentiated in two layers (Hoffmann 1985): horizontally as a differentiation from sub-disciplines and sub-fields, vertically within a discipline, starting out from an abstract scientific language to the concrete factory and end-user language (overview on the different approaches in Roelcke 2005: 38–42). Within the science of terminology, the combination of knowledge and language constitutes an independent subject with own models for theory and practice (Felber and Budin 1989; Budin 1996; Arntz, Picht and Mayer 2004). Thereby research is mainly interested in the construction of knowledge systems and their transfer (interlingual and intralingual, see Gerzymisch-Arbogast 1996; Jahr 1993; Lundquist and Jarvella 2000), whereas practice is rather concerned with questions of management (Wright and Budin 1997/2001; Ibekwe-SanJuan 2007; see also International Network for Terminology, www.termnet.org) or with individual questions regarding the handling of concepts and denominations. This clear distinction in theory leads to problems in practice which are on the one hand related to the language awareness and comprehension of the communication participants and on the other hand to historically grown language usage in the several departments of companies. Different from questions of knowledge production in research, such communities of practice (see Pogner in this volume: Pogner 2012) involve strategies of action facilitation. Different positions are in opposition to each other here: DIN 2342 (2004), for example, sets separate norms for the concept formation. Some supporters of terminological applications demand a strict 1:1 allocation of concept and denomination (e.g., Göpferich and Schmitt 1996), i.e., the absolute avoidance of synonyms which are common practice in in-company communication, however, causing confusion in external communication (e.g., in instruction manuals). Villiger (2008) on the other hand, points out that the usage of different synonyms for one concept in a company can indeed make sense if different options for action in individual departments are signalled thereby. She also indicates that in practice the Know how in a company due to the terminological systems of concepts mainly remains unused.

3.3 Models and reality

A rather indirect convergence between theory and practice can be stated if knowledge in the form of models is taken as a basis. Johnson-Laird (1983) introduced the concept of the "mental model" into the scientific discussion. Even though the term has remained relatively vague (or just for that reason) it constitutes a useful instrument to emphasize the essential aspects of an object.

The relation of the analogies is substantial, i.e., only individual features of the model correspond to the original. Models are employed to describe relations which are otherwise not directly accessible. From this point of view they also function as comprehension patterns. They indicate how an object works or is handled, how a concept or a situation are to be understood (Dutke 1994).

Let us look at the example of an artificial leg. The construction of this object contains a model of human locomotion. Otherwise an application would not make sense. The function of playground and sports equipments is based on corresponding models of playing or the respective form of sport. Driver assistance systems simulate in the application certain driving situations on the road. The modelling of situations and scenarios is a popular research instrument. In practice, models are given directive character for the reproduction of artefacts or as an instruction for their usage. With regard to the technical documentation one can differentiate between two perspectives which model the usage of technical products: product orientation and user orientation (Reinicke 2004). The former focuses on the object with its features and functionalities. The role of the user is restricted to the operation of the appliance according to its construction. The latter focuses on the users' objectives which are pursued employing the device. The description involves actions which lead to the particular objective or actions which have to be avoided since they would endanger it.

3.4 Schemes and understanding

Whereas models as structural models reduce the complexity of the excerpt of reality, schemes include the diversity of combinatorics. The scheme theory (overview in Schnotz 1994; see also Konerding 1993; Ziem 2008) assumes that we perceive the world and act in it according to certain standards of experience and expectations. The field of computer-based knowledge processing has contributed the description in form of attribute-value pairs (or slot-filler structure) which has become accepted in many domains for the capture of relevant knowledge. Accordingly, a scheme is a form of organization including general categories in a clustered form which are specified for an individual case. Thus one and the same scheme can be applied to many similar cases. This procedure is well-known from standard forms. There is, for example, a position with a general category (attribute), such as the name, which is individually allocated a value (in the example the individual name). Categories structure a subject in a way as it appears to be relevant for us, e.g., in everyday life or in profession. This process captures all possible semantic relations which are by default part of understanding a concept. Thus we grasp objects according to their spatial extent of length, width, height or object conditions according to measurement categories such as temperature, weight, speed and so forth. The compilation of categories within the schemes is variable. They can in principle be extended indefinitely; the embedding of a new scheme in an existing category enables every degree of complexity. In the application they allow the specification of the general. Due to the individual value assignment they serve as an instruction and finally, with regard to the general categories, facilitate the comparability of objects and facts. They are applicable to static objects (frames), to events or processes (scripts) as well as to action patterns (plans). In textual linguistics they are relevant for the analysis of the knowledge thematized in the respective text (already Metzing 1980; Stenschke and Wichter 2009) or vice versa, as structuring patterns of the knowledge inventories which are researched for the text production (Eigler et al. 1997).

Understanding, however, also involves problems which are related to our cognitive conditions. For this purpose let us look at the origin of the scheme theory. It goes back to cognitive psychology according to which our perception is influenced by our knowledge, i.e., by learnt schemes (Bartlett 1932). One consequence becomes apparent, e.g., in the process of writing or reading texts where we understand considerably more than what is thematized in the text. We add knowledge which we already have or we assign what is being written or read to existing schemes and deduce our mostly implicit conclusions (Rickheit 1991). Insofar as humans learn different things and have different experience, they possess different knowledge schemes. This means that they interpret utterances about objects and facts in different ways and draw different inferences. Nevertheless certain (partial) conformities can be assumed insofar as humans act in culturally determined environments which are characterized by particular and definable systems of knowledge, values and actions. Communication across these respective cultural shapings must therefore be regarded as a specific communication task (Rothkegel 2008). Thus one can distinguish between communication among experts and communication between experts and non-experts. Taking into account the strong differentiation of the disciplines, the main part of communicative events might belong to the latter form. Thereby the attributions "expert" or "non-expert" are to be understood as changing roles regarding the respective level of knowledge in reference to a subject. Additionally, two further differentiations are of relevance: the distinction between knowledge, ignorance and non-knowledge and the distinction between specialist knowledge and everyday knowledge.

The schematic description of knowledge (knowledge representation) is particularly suitable for distinguishing between ignorance and non-knowledge

(Geisenhanslüke and Rott 2008). Non-knowledge suggests that there is the awareness about the existence of a scheme of a knowledge inventory which one does not know. The corresponding information, however, could be gathered. The didactic problem of how to generate non-knowledge is to be understood in this sense. Ignorance, on the other hand, means that such awareness does not exist, i.e., one does not know that one knows nothing. Communication problems can also include that communication partners know something different or something wrong without becoming aware of it. Such situations are typical for the expert/non-expert communication if, for example, everyday knowledge is seen as specialist knowledge and vice versa. The instrument of schematic description permits the creation of transparency in communication. This option is, however, of theoretical kind. In practice, schemes are primarily applied for the description of action-relevant specifications of objects or facts.

4 Actions and action spaces

4.1 Physical and communicative actions

Human-technology interaction takes place in action spaces which involve actors, systems and persons concerned. These action spaces are in turn communicatively revealed by the actors, now in the roles of writers and readers or users. The contents of communication refer to the knowledge inventories about objects and facts which are specifically subjected to changing conditions. In action-reaction cycles, actors and systems collaborate with regard to a task and interact with regard to their own dispositions. Thus the changes do not only refer to the objects being worked on but also to those involved in the process themselves. Concerning the actors the focus is laid on the cognitive-communicative provision of skills, attention, discretionary competence, decision-making and responsibility for action. With respect to the systems, importance is, amongst others, attached to start and end functions as well as control and protection mechanisms. The communicative control is effected through the illocutions expressed with the utterances. They set the subject matter, i.e., the propositional content in relation to the actors and facts. According to the speech act theory (Searle 1969) and the linguistic adoptions (e.g., Ehlich 2007; Staffeldt 2008) as well as the practical applications (see as functional design Muthig and Schäflein-Armbruster 2009) a distinction is made between declarative statements, claims, descriptions as representing illocutions, or commands, recommendations and warnings as appealing illocutions. Theory mainly concerns questions of the specification of the components (fundamental Rehbein

1977) and the identification of patterns and combinatorics in texts (Schröder 2003) and dialogs (Weigand 2004). Practice is interested in formulations and specific markings which avoid possible misinterpretations. A suitable action model integrating both aspects, interaction and communication, is important for further research. With regard to an operational approach, action logic provides an instrument for bridging both (von Wright 1968).

4.2 The structure of actions and semantic roles

The approach of action logic systematically connects states, events and actions. Events are regarded as changes of states, actions as intentionally caused events. The deduction of actions from states and events facilitates a precise description of changes of object states and thus the identification of individual phases in technical processes. Pursuing this approach at the sentence or proposition level, one arrives at the semantic roles. They mark the components which structure actions and reactions in interactions of actors and systems into certain configurations. The deep-case theory assumes that all actions can be captured as specified configurations of such semantic roles (de Beaugrande and Dressler 1981, 53–87). Examples are ACTORS (causer of an action), OBJECT (focus of an action), INSTRUMENT (medium for the execution of an action) and so forth. In a practice-related version of grammar books (Pörings and Schmitz 2003) and writing guides (Weissgerber 2010), semantic roles appear in a conventionalized way in form of w-questions.

Actions are structured in themselves. A distinction is made between three structural patterns: complex actions with partial actions, action levels with objective, strategy and execution as well as action phases with planning, execution and control. Examples for partial actions, which together form a superordinate action, can for example be found in production processes. From a semantic perspective it is interesting here that the thematized object states refer to new referents (effected objects) from one step to the next and that nevertheless an object continuity is understood within the reference range (see analysis by Rothkegel 2010: 43–44). The segmentation into partial actions or work steps allows the centralization of very different situations and individual scenarios where all involved parts, with the exception of the object, can be exchanged.

The distinction of action levels permits the description and instruction to be differently granulated depending on the action competences of the respective actors. Thus it may be sufficient to phrase: *save file* (objective = file is saved), *copy file* (functionality offered by the tool as a strategy of saving) or *click once "save" or "save as" in the menu*, whereby the execution is indicated. This involves only one action which is communicated through so called by-actions: A file is saved by copying it, by clicking "save".

The distinction of action phases aims at target states and actual states of the involved objects. Thereby the actors perform different roles. The planning of actions includes that they are documented as action plan, guideline or concrete instruction in the way that others can put them into practice or assess them. Apart from information about objective, strategy and execution there are initial and final state as well as prerequisites, conditions, risks through possible dangers or errors and consequences to be considered. Furthermore it is important to define alternatives and decision points. All parameters can be object of checking the actual state, i.e., of the result object or the actually running processes.

4.3 Modelling tasks

The differentiation of action components allows a description with concrete, if necessary quantifiable information as they are required for practice. This particularly applies to the use of tools. They imply a model of effected or modified objects or a model of working processes. Both appear combined in the tasks the tool is constructed for. We exemplify this by means of computer or software in their tool function. There occurs a dynamic relation between actor, task and system (Rothkegel 2010: 59–60). In a similar way Hackos and Redish (2000: 55–75) combine their approach of task analysis with the users' work at their workplace. The actor on the one hand defines an objective which determines the decision on the employment of a system with regard to strategy (functionality) and execution. On the other hand his intention is directed towards decisions that change the initial state of his data in the direction of the final state.

The following illustration exemplifies how an actor manipulates a relatively simple task with regard to a system. The objective of an author could be to produce an abstract of 300 words. He uses word processing software which supports the typing and layout work as well as the check of the length of the text being created. The author's purpose is preferably not to produce an unnecessary text length. Thus he defines the initial text state for the check at a length which he assesses as approximately the half. By comparison with the specification delivered by the system, he receives the information on how much text remains to be written. During the further writing process he may repeatedly do such a check and eventually delete some words until the comparison yields 300 words. The author can utilize the system support by breaking down the task into individual steps with iterative loops and completing the execution according to the system defaults. This is his contribution to the solution of the task. The contribution of the system involves offering appropriate individual steps in form of functionalities and execution options through the user interface.

Software as a technical product is prototypical for the phenomena of human-machine interaction (Geiser 1998; Heinecke 2004; Dahm 2006; see also article by Schmidt in this volume: Schmidt 2012). Particularly the feature of interactivity makes the product interesting for a user-related orientation when using technology. The tendency towards assistance systems (user assistance) is also to be understood in this direction. A complete takeover of tasks and fully automatic operations are replaced by division of work and specific help if necessary leaving the actors scopes for decision-making and, at the same time, relieving them from time-consuming and inconvenient partial steps (e.g., counting words). Application areas of this kind mainly include support systems for cooperative working which are gaining increasing importance in the professional field. Examples for assistance services are the documentation of work progress, the dynamic adjustment of the sequence planning during the completion of projects, the availability of messages and documents or the monitoring of deadlines.

During the interaction the system comments or reacts on the users' actions. The comments include success messages, warning messages and error messages. However, fully automatic systems, e.g., driver assistance systems, also allow for autonomous (and amongst drivers partly controversial) interferences into the actors' activities. But let us first of all consider computers and user interfaces as the meeting point of technology and language usage. There are, for example, requirements for user-friendly system messages which are formulated out of practice and set as a norm (DIN EN ISO 9241): The imparted information is supposed to enable the actor to control the dialogue, spot errors and determine further action steps which includes that he is able to clearly identify the functions triggerable from the message window and the push button. The messages should be comprehensible with regard to terminology and style. "Wording" describes formulations the users are confronted with on the screen. Apart from space problems which have to be solved first by the software engineers and later by the translators during the localization, focus is laid on the correctness and consistency of the surface terminology which support or hinder the users in their search and selection decisions. This problem can, however, not be solved technically but rather demands research into the question of how users actually deal with interfaces and what problem solving strategies are employed by them (Wagner 2002; proposals in Felix and Uhr 2000; Haase 2005; Jacko 2007).

4.4 Action and reaction

Based on the context, the illocutions are of specific importance. With them the actors are placed special emphasis on. In communication contents are not merely transported but simultaneously transformed with reference to the actors. Facts become a statement, message or explanation, a recommendation or warning, a promise of guarantee and safety on the part of the manufacturers, a demand or obligation of the consumers. A complex text becomes a report, an instruction or argumentation. The problem, however, is that this cannot immediately be seen in the utterance or the text in the way one can recognize the structure into chapters or the typography. Furthermore it cannot be understood in the way one can understand the content or the thematized knowledge. Illocutions can only become accessible as interpretations, i.e., by recognizing or analyzing the conventions according to which they are encoded. This includes certain indicators, cotext and context. Literature on speech acts and speech act theory has therefore developed models and methods which facilitate the identification and classification of speech acts in different contexts (e.g., Ehlich 2007; Staffeldt 2008).

The communicative mediation of action-reaction cycles in the human-technology interaction involves a number of different actors: technology producers, technology users, decision makers, law- and norm-makers, marketing specialist, technical authors and readers. Accordingly the influence mutually exerted by the communication participants is of different kind and extent. The range of speech acts corresponds to this. Orienting towards the five classes of speech acts which, according to Searle (1969), consider both the interpersonal interaction between speaker's disposition (writer's disposition, S) and listener's disposition (reader's disposition, H) as well as the relation between content and state of the world, the core aspect of the change of states, as we assumed it for the characteristic of physical actions, can also be applied for communicative actions. In this sense the disposition of speaker and listener describes the respectively intended impact on each other. Thereby S is to be seen in the proactive role, H in the reactive role. Both the initiating action and the reaction lead the continuation of the interaction into a particular direction. Let us illustrate this phenomenon with two examples.

In the case of representatives ("representation" in accordance with Bühler 1934), S signifies to H, which facts he/she (S) regards as true, certain, possible, necessary, useful and valuable, etc. With reference to human-technology interaction this concerns, for example, descriptions (*This armature is equipped with a flow limiting device.*), evaluations (*Twelve current navigation systems were tested: none was perfect.*), definitions (*X is a editing system for the modern generation, administration and production of technical documents.*). H's reaction can be positive in the way that he understands the information as an action enablement, i.e., he includes the flow limiting device in the course of the installation actions, does not purchase any of the navigation systems and considers using X. In case of a negative reaction, a possible action disposition is rejected.

Directives ("appeal" in accordance with Bühler 1934) directly aim at action dispositions of listeners, readers or users. From a theoretical point of view it is not completely clear whether instructions are to be understood as real demands or if they are rather to be assigned to the descriptive classes. The use of imperatives on the expression side alone is not decisive; it is, however, regarded as a simplification of the perspective. In the classification of the types of thematic expansion, which Brinker (2005) has based on the speech act classes, they rather have a descriptive character. An aspect in favour is the property of directives to leave the communication partner the possibility for rejection even if this may be connected with sanctions. Insofar reading requests can be understood as a real directive since sanctions are concurrently stated here: *Please read carefully the information in the user manual before using the appliance. [...] The manufacturer is not liable if the information is not followed*. This is different in the case of instructions such as: *Turn the temperature control to the desired position*. A deliberate rejection of the action on the part of the user would be absurd in this case since it can be assumed that this information is desired by the user in order to attain his objective. In this sense, instructions are to be regarded as descriptions of action plans.

4.5 Evaluation of the situation: safety and security

What is interesting in practice, e.g., in case of knowledge transfer through technical documentations, is not primarily the classification of the illocutions but rather a precise evaluation of the situation with regard to actions and reactions in the interaction between actor and system. The actors are to be clearly informed about which changes of state are to be effected by them according to the action plan and which ones will be caused by the system. Moreover they need information about the states of the objects worked on. Such situation evaluations are particularly important if possible dangerous situations are involved. In practice it is tried to standardize such situation descriptions as safety/security instructions – even if mainly on liability accounts. Propositional and illocutive level are blended here.

An analysis of the pertinent norms (ISO 3864; EN 61310; EN80416 or the US-American ANSI norms of the series Z535) shows that they are based on different models which partly overlap, though often in a fragmentary way, which is criticized by practicians themselves (Bauer 2000; Rothkegel 2010: 66–70). A common basis, however, can be seen in the issue of attentiveness. Signal words, graduated according to the risk and supported by pictograms, are employed: DANGER, WARNING, CAUTION or ATTENTION. Now risk is a theoretical concept describing the calculable relation between the probability of an accident event and the possible degree of damage. Given the same degree of damage (death), DANGER and WARNING refer to different probabilities. DANGER stands for imminent danger (e.g., in the operating range of a robot), WARNING for possibly dangerous situations (e.g., explosion through dangerous substances). Further differences in the degree of damage are expressed through CAUTION and ATTENTION (minor injuries, material damages). Apart from the attentiveness towards the source of danger and type of danger, the following specifications are required in order to realize the situation: possible consequences and measures of danger prevention. A distinction of prevention measures against a source of danger (e.g., technical containment of a dangerous substance) and safety measures for persons and objects that can be affected by the damage (e.g., ventilating the room in case of leaking dangerous substances or leaving the room) is not marked although this would be important for the situation evaluation and action dispositions on the part of the users. In the former instance it would be about control measures, the latter case would concern rescue measures. The example is supposed to clarify to what extent a conceptual, scientific modelling (e.g., action concept) can be relevant for practice (Schmitt 1995; Villiger 2007).

5 Texts as mediation forms

5.1 Theories of text

Since the 1980s the text concept has been a subject of modern text linguistics (de Beaugrande and Dressler 1981; van Dijk 1997; Heinemann and Heinemann 2002; Adamzik 2004; Janich 2008). With the pragmatic turn and the focus of research on linguistic usage, the text had become the communicative base unit. This refers to spoken as well as dialogic and written texts. The determination of text characteristics was the central interest of theory and method development. Text models and analysis methods examining regularities and systematic in texts were gaining increasing importance (Brinker 2005; Carter 2006). The mass use of electronic media brought into focus questions about their impact on the text as a communication form (Kitzmann 2006). In research automatic text analysis has been developed (Köhler 2007). At the same time software of text processing became a self-evident tool of textual work. In professional practice the processes of creating documents have increasingly been supported by authoring and editing systems (Ament 2003; Closs 2007). They are supplemented by management systems which manage documents (Götzer et al. 2004) and individual contents (Hackos 2002; Rothfuss 2002). The perspective of analysis with questions like "What is the case, what happens?" in the linguistic disciplines is opposite to the perspective of production in practice. The latter on the one hand demands comprehensible writing and on the other hand a cost-effective production which is to take place in a standardized and, as far as possible, automatic way. It is to be mentioned that simultaneously and partly in mutual influence of cognitive psychology and textual linguistics, scientific interest in text production and writing processes had developed. Whereas linguistics focuses on the text and how it is produced, the interest of writing research is placed on the writers and their activities. The increasing relevance of writing in the professions has entailed the development of a counselling culture which includes many individual aspects from all these directions between theory and practice (Gerson and Gerson 2006).

Against this background of mixed approaches we focus on the conceptual field of comprehension, understanding, and intelligibility. This is of central importance for both, theory and practice when dealing with texts. Let us set out from the already existing text which we want to understand. Theory and practice use completely different approaches here. In a nutshell, theory concerns the "understanding of", practice the "understanding for". The former deals with analysis and instruments of analysis, the latter expects action competence to be imparted by the text. Thus questions about knowledge, themes and situational parameters are opposite to possible usability and ergonomy tests.

Theory distinguishes between text comprehension and text intelligibility. Thus the cognitive processes running in the readers are separated from the characteristics of the text which ignites these processes. The interest of linguistics in the text as a complex but autonomous structure poses the question of what a text yields in a linguistic sense. This, however, depends on the text models which the analyses are based on. Despite all differentiation of the distinct components, they share an integrative point of view. Unlike in case of the employment of management tools in practice, which require a perspective that principally separates content and form, text modeling aims at pointing out the links and references which make the text an appropriate organization form for communicative purposes. The seven criteria of "textuality" by de Beaugrande and Dressler (1981: 50–215) which identify a linguistic structure as a text, provide an inventory of accesses for the analysis of the text (cohesion, coherence, intentionality, acceptability, informativity, situationality, intertextuality). Cognitive and pragmatic aspects merge together here, text surface and text depth are differentiated but related to each other. The lexis, syntax and textual organization realized at the surface reflect what the text was produced for. Our approach focuses on the four criteria related to content and function. Knowledge relations result from coherence, i.e., due to the thematized concepts and facts and the consequently available potential of inferences in the cognitive work of the text actors. This invisible part is assigned to the text depth. The grammatical relation at the text surface, which expresses the references and the relations between them, is specified in the characteristics of cohesion. The interaction between producer and recipient is involved in the criteria of intentionality and acceptability. Within this frame, text comprehension is directed towards the WHAT: knowledge, information, theme.

5.2 Levels of text: knowledge, topic, function

Knowledge is not to be equated with text or information or even with theme as it may well occur in practice (Rothkegel 1993; Jahr 2007). Being a communicative unit, the text translates selected knowledge inventories into themes which, according to Brinker (2005), are functionally unfolded. Considering a theme as a question, which the text supplies answers for, the theme provides a criterion for the selection of individual knowledge inventories which are considered as information being sporadically distributed in the text. The selection and distribution on the other hand depend on the text function, i.e., on the author's intended impact on the reader. Brinker (2005) distinguishes text functions in parallel to the five speech act classes introduced by Searle. Thus it can be consistently assumed and possibly explained that an information in an argumentation is of different importance than in a narrative text or that a term in a descriptive environment forms a different conceptual field than in an instructive or legally defined context. In this respect we distinguish between knowledge and thematic concepts or key concepts.

Even though knowledge is regarded to be linked to a domain, it is independent from its application in the individual text. However, it can for example be of interest for us which knowledge inventories from which domains can be found in the text to be analyzed. We assume the coherence of the text and bring in our knowledge as far as possible. We look for expressions which we can attribute to specific domains and internally arrange the groups according to hierarchical orders or in schemes, as outlined in section 3. If the expressions do not describe generalizations themselves, they are thereby assigned to superordinate concepts or general fields of knowledge. Schemes following the principle of co-occurrence of knowledge inventories are particularly suitable for capturing situations which are conceptually elaborated in the text. Such analyses may provide information on how far the knowledge thematized in the text is to be classified as technical or general-language. This is possible with regard to different configurations of the attributes and/or different specifications of the corresponding values. It shall be illustrated by the short example of the concept of *risk* with different attributes:

natural science domain:
risk *(probability of a disaster = [...], degree of damage = [...])*
financial domain:
risk *(possible chance = [...], possible loss = [...])*
every day language:
risk *(danger = [...])*

This is different in case of expressions which represent the theme. Here the aim is to know which conceptual containment with reference to a certain question is made and which role a conceptual field plays in an argumentation or instruction. A special role is played by key concepts. They connect the several knowledge pieces of a domain which is independent from the text and are regarded as super-ordinate concepts for the themes elaborated in the text. But let us remain with the theme-relevant concepts themselves. The concept of *sustainability* can, for example, refer to advertising for technical solutions in terms of a reduced CO_2 emission or to action instructions in terms of the disposal or reusability of appliances and packaging materials as it is common in instructions for use. The analysis thus focuses on the relevant expressions in the text which are pooled according to the theme (or the themes). They represent the respectively thematized point of view. The organization and internal grouping of such conceptual fields can be based on two strategies: macrostructures and mind maps.

5.3 Structures of text

From the point of view of linguistics, macrostructures might be regarded as prototypes for textual structures in general. Introduced by van Dijk (1980), they constitute an instrument of text structuring covering the whole text. This method is inspired by the assumption of a separation of text surface and text depth similar to the distinction in transformational grammar which relates to the sentence. Accordingly, van Dijk tried to formulate rules (macro rules) for the transformation of a running text, segmented into proposition, into a hierarchical thematic structure which can finally be presented as a tree structure with knots and edges. Even though the concrete application of these rules (selecting, deleting, generalising, constructing) proved to be inconvenient and less feasible, they represent useful strategies for the generation of a graphically supported text representation. Macro-analysis utilizes the formation of units (paragraphs) in the text as well as headlines or the possibility to construct them (Stein 2003). It is also suitable for structured representations of action plans which break down the complex action into partial actions or components. Insofar as macrostructures pool central information in the respective communicative view, they also qualify as good planning instruments. This is in turn facilitated in the way that the planning in terms of formal structure can be supported by standard software.

This also applies to mind maps which likewise refer to thematic relations but have become more prevalent in practice than macrostructures (AVIDES Media 2005). The structuring of contents is defined less strictly. It is about the grouping of concepts around a centre or various centres, whereby the relation between the concepts is of merely associative nature. As a rule, concepts which are placed closer to the centre have a closer semantic relation to the central concept than those placed in the periphery. Mind maps have become important for brainstorming in groups and the task to compile components of a theme. Reciprocally, the method can also be employed for analyses. In that case the questions to the text aim at determining which expressions belong together and thus form a kind of concentration towards the core. Analyses of proximity and distance can be used to ascertain which concepts form themes or partial themes, i.e., function as key concepts.

Let us come back to the above introduced comparison of "understanding of" and "understanding for". The use of taxonomies, schemes, macrostructures and mind maps tends to an "understanding of". The interest, related to analysis or production, is directed towards the comprehensible and repeatable perception and classification of objects and facts or the models imparted in this regard. These can be more or less correct, complete and explicit. This also includes the captured contexts, i.e., situational reference, media dependence, social classification. Classifications, specifications and thematic elaboration generate a reference range in which the actors mentally move and which they can mutually refer to in the "discourse community" (see Pogner in this volume: Pogner 2012). "Understanding for" on the other hand aims at action competence. The readers are replaced by the users here. In place of precision and completeness of mental models focus is laid on usability and ergonomy. It concerns practical knowledge (Hörning 2001) which is again imparted by the text. Compared to technical products, the text likewise becomes a product which can be of different use for certain purposes or tasks, e.g., useful for the employment and execution of a human-machine interaction.

5.4 Perspective of the users

Usability is tested for both a text and a machine. The application is simulated similar to a test run. Test persons perform a task on an appliance leaning on a corresponding documentation fragment. It is observed whether and how and in which time the set task is completed (e.g., setting a digital alarm clock). It can now be asked if the setting may not have been completely accomplished due to problems during the performance or because of a technical term not being understood or for the reason that the mental model of the space as set up by the actor, did not correspond with reality. This should, however, be differentiated. Even if the solution was successful it cannot be assumed in any case that the text was all right. The approach here is to supplement the review of the result and the measurement and interpretation of pauses by methods of the "think-aloud protocol" or subsequent interview (Hennig and Tjarks-Sobhani 1999, 2007). They can all indicate flaws in the text and suggest possibilities to optimize the intelligibility. This is rather a side-effect which is, however, gladly used. Practice, in contrast, attaches priority to other aspects of usability: effectiveness, efficiency and user satisfaction (see DIN EN ISO 9241, 11; Nielsen (1996: 275)) points out five criteria for the usability of websites: easy to learn, efficient in use, easy to keep in mind, low error rate, compliant in use. The review of texts can relate to very different dimensions of expenditure. This reaches from proofreading to checking the factual correctness and completeness, from checklists to specific error analyses and finally to tests associated with the appliance tests in test laboratories specifically set up for this purpose. So far these facets have hardly been regarded in the text models of established text theories. It remains to be considered to what extent such a "practical" view (in accordance with Hörning 2001) on the text can change the theoretical concept. Concerning this, the concept of quality could build a bridge.

5.5 Quality of text

With reference to technical products, practice distinguishes between product quality and process quality. Quality standards are for both kinds set through requirements. They are checked by means of procedures of quality assurance which contrast actual values and target values. The question is whether such principles can be transferred to texts and how text quality can finally be determined (Göpferich 2004)?

Let us first focus on requirements directed towards the text as an existing object. Then we will approach process quality whereby the text is evaluated with regard to its production process. Considering the text characteristics we arrive at graduations which can for example be identified through the realized text structures on the basis of coherence or cohesion. Thus clear macrostructures, i.e., clear and stable forward and back-references, set a benchmark for the quality of a text. It is possible to employ an evaluative comparison of macrostructures to ascertain which text is better in comparison to another. This also takes into account that different readers, due to their diverse backgrounds of knowledge and experience, come to different structures which correspond to their respective mental models. However, analyses of this kind require a textual linguistic competence. Other practice-related approaches rely on linguistically neutral procedures such as measurability and countability, e.g., with regard to readability (Hennig and Tjarks-Sobhani 2000). The reading speed is an important measurement parameter here. Decisive are word length (number of syllables), sentence length (number of words in a sentence), the amount of monosyllabic and polysyllabic words as well as the wide-spread use or scarcity of words in language usage, which is attested by frequency lists. The readability can be measured on the basis of a readability index, which is captured through a readability formula. This procedure is, however, critically challenged with regard to its validity, particularly if the thus calculated readability is related to or even equated with the evaluation of intelligibility.

The basic assumption that conciseness in terms of word, sentence or text length is decisive for the text intelligibility is to be critically assessed. What is disregarded here is the aspect that the parameters of conciseness and precision are often in conflict. Shortenings are usually related to incompleteness, inaccuracy and generalization. It is though comprehensible that everyday practice relies on countable parameters of this kind due to the requirements of instructions. This, however, already concerns the text production i.e., possibly simple recipes of generating intelligibility. An example is the method of "Information Mapping (IMAP)", patented in the USA, which links the structuring of school knowledge (e.g., per text unit one theme which can be formulated as a headline) with the results of a scientific memory experiment (summary in Böhler 2008: 145–163). The recipe-like portioning and quantification of information based on the "magic number seven +/- two" is easy to be implemented. However, the question is to what extent the restriction to units with seven countable elements (words with seven letters, conceptual fields with seven concepts etc.), which in Miller's experiment (Miller 1956) was ascertained to be favorable for the retention performance in short-term memory, can basically be transferred to the comprehension performance in texts which is about the formation of coherence and new contexts (also see criticism in Jansen 2002). A global, quantified determination meets by no means the requirements of the communicative situation.

Quality can also be determined in terms of the product's production process. Process quality is based on the assumption that the correct execution of the planned partial processes (target values) ensures the quality of the resulting product. In this sense the technical documentation as product is opposite to that as a process. Due to the variety of specific relations and contextual dependencies, a product is difficult to standardize. Production processes, on the other hand, can be divided into similar procedures, phases and components. This allows a standardization and modularization. Principles of organization and management apply regardless of the contents and communicative purposes of the documentation as a product. Thus Grünwied (2007: 50–51) structures software documentation into five phases which can likewise be applied to documentations of all other technical products: 1. planning with parameters, 2. analysis and conception with target group and operation analysis as well as editorial guidelines, 3. creation and correction with drafts and final version, 4. production and distribution including the definition of media products and finally 5. evaluation and update with test reviews and change requests. As common in guides, the bullet points are very generally phrased; concretions are illustrated in form of examples and counter-examples (e.g., Juhl 2005).

5.6 Iceberg-model of text

The separation of content and form constitutes the basis for the use of software tools. Their application requires the possibilities of a formal comparison. Word, sentence and paragraph boundaries are identifiable, likewise are positions in the sequence, distinctions of beginning and end as well as the correspondence of similar character sequences. Texts can be manipulated as data by corresponding format specifications or other annotations. The question is to what extent the formal characteristics represent the content, which is what we are actually interested in? In the context of the iceberg model of the text this is conceivable – at least to a certain extent (Rothkegel 2010: 123). Accordingly, the structure of the content is reflected in the visible surface of the text. This can, however, only be assumed if the text design allows for this structure. In principle, the same is true for the reader. We gain access to the content through that which we see or hear from the text. It is the text surface that catches our eye with its words and graphical material, with the distribution of colored units and white space on the page, with blocked paragraphs and typographic normalities and peculiarities. This is the visible "tip of the iceberg". As formal unit it builds a bridge to the addressees, whose attention is to be aroused and maintained. The part in the text depth, on the other hand, includes knowledge, themes, functions, implicit messages with regard to the relation between producers and recipients. If such information is to become manipulable data, it has to be added to the text through appropriate comparative operations with wordlists or as metaknowledge in the form of annotations to the text, as in case of the indexation of texts (Villiger 2008).

The considerations show that texts in the theoretical sense and texts in the practical sense are two different matters. Nevertheless there are interfaces. The questions asked in practice can be transformed into theoretical questions. For example: What analytical instruments are suitable for the determination of text quality in terms of creating action competence in advance rather than through tests afterwards? This also applies to the relation of action competence to intelligibility and attractiveness, reference to addressee and acceptability, usability and situationality. On the other hand, it can for example be asked to what extent concepts like coherence and cohesion can be useful for the determination of effectiveness and efficiency in terms of text usage.

6 Cultural aspects

Examinations into culture-specific aspects of communication on technology are scattered throughout different disciplines or sub-disciplines, each dealing with communication, texts and cultures from their own points of view. On the other hand, requirements from practice provoke various insecurities regarding the international handling of technical documentations. This is particularly true for the field of communication on safety which is on the one hand important with regard to the users' action competence and on the other hand characterized by national norms. Independent of these norms, questions arise with regard to the "self-evident" (Hansen 2011) or that which is not to be assumed as self-evident in the use of technology and text reception in different cultural spheres. Two dimensions, interlocked with each other, become important here. Thus it is about the cultural specialty of technology itself as well as the corresponding handling of it. In this sense it is referred to a technological culture (Hubig and Poser 2007; Rösch 2008; Banse and Grunwald 2010). It is obvious that the technological culture influences the kind of documentation, i.e., the way technology and its use are described and evaluated. Furthermore also the text specialties themselves are subject to cultural influences. Let us illustrate this on some examples.

Development and handling of technology are per se linked to society and are thus regarded as cultural practice (e.g., Larnvik 1996; Irrgang 2005). There are noticeable country-specific constructions and realizations such as electrical appliances with (e.g., Germany) and without (e.g., England) power plug, right-hand traffic or left-hand traffic with the consequence of a reverse polarity of the electrical installations or also in the form of spherical (European) or cylindrical (American) reactor buildings. As an example, Schmitt (1999: 113) mentions the thematization of the Italian product design as "Haute Couture in sheet metal" against the stressing of German technology. What is also remarkable here is the aesthetic consideration, which Rogge (1980) summarizes in the headline of an article in the following way: "An engine must look like a birthday present."

The culture-specific aspects of texts are related to communicative habits of different discourse communities (Hoft 1997; Min-Sun 2002; Hayhoe 2006; Lum 2006; Rothkegel 2007). There are, for example, differences with regard to problem handling in project management (Ulijn and Strother 1995; Poncini 2004) in the way that in the French context a problem area is rather generated from different points, whereas in the German context a sequential approach of individual problem descriptions and solutions is preferred. Also written texts show culturally determined peculiarities. Thus, Gutermann (1996) compares the different spatial representations in German and French technical texts. Barnum and Li (2006) go further into the question to what extent American and Chinese technical documentations differ in terms of their production process, their reception and use. This reveals similarities and differences with regard to historical and economic influences as well as educational influences. Text structures likewise prove to be culturally determined and with them the respective mode of intelligibility (McCool 2006). Last but not least it must also be assumed that terminologies are not independent of an individual language but, similar to everyday vocabularies, are based on culturally determined conceptual systems.

7 Outlook and research questions

7.1 Interfaces between theory and practice

To sum up we outline a number of research questions which may result from the above discussion with regard to the interfaces of theory and practice. Such questions are to be seen in the context of a pragmatically founded communication model, as described in section 2. We consider how the thematic fields of knowledge, action, text and cultural aspects can be elaborated in a perspective of communication on technology that integrates theory and practice. In a rough differentiation we contrast the scientific approach of modeling, structural representation and conceptual classification systems with practical requirements to create action competence. Analyzing, grasping, explaining and describing on the one side compete against activities of an effective and efficient production in planning and execution on the other side.

7.2 Knowledge in use

Dealing with knowledge often leads to communication problems in practice. It is difficult to realize the fact that humans understand different things even if phrasings and mediation strategies try to avoid this. What is often lacking is transparency of what is similar and what is different. The crucial question is whether and how this transparency can be created with linguistic means? This includes specific sub-questions like: What analytical methods can be employed to develop the conceptual systems which allow for the situational references (text analysis, questionnaire, interview)? To what extent must themes, thematic structures, thematic unfolding and perspectives be considered to facilitate conclusions about the criteria of relevant knowledge selection (dependence of knowledge structure on theme and text function)? How are technical terminologies of different domains linked to each other in the individual text or conversation? What is the relation between comprehension and acceptability in the expert/non-expert communication? Can the same analytical methods be employed for the cultural comparison?

7.3 Action competence

In the first instance it may appear self-evident that transparency is a prerequisite for the creation of action competence. It is important for the "understanding of". For the "understanding for" it may be secondary. Here, completeness, identifiability of the reference ranges and clarity of the task descriptions are of primary importance. What scientific instruments can be of practical relevance in this regard? Did the dealing with knowledge involve conceptual questions; do questions of modeling come into focus? How can human-technology interaction be broken down into useful components which clearly describe the respective action-reaction cycles with the roles of actor and machine/system? Which (systemic) models from the field of technology are compatible with interaction models of the speech act theory, conversation analysis or dialogue analysis? Can a dialogic character of this interaction be assumed? Is there a parallelism between physical and communicative interaction? How do changes of the interaction, e.g., an increased use of autonomous assistance systems, affect the communication on technology? What do the users have to know, are they included and do they want to be included at all? Above all, there continues to be a lack of empirical research and a corresponding research design in this regard.

7.4 Communicative competence

Talking or writing professionally about technology requires communicative competence. What is part of it, which components are important and how are they related to each other? Are the models of textual linguistics sufficient to bring together text production and reception with practical criteria of quality, i.e., product and process quality? What is the role of technology-mediated communication, what kind of communication is predetermined in the models of communication technology? How do changes of the spatiotemporal parameters, determined by them, affect the interpersonal communication? Does the obligation to inform implicate an obligation to collect information on the part of the users? What consequence does such a tendency have for the barrier-free accessibility to media? Does research, apart from analysis, permit reliable forecasts?

7.5 Conclusion

The applied sciences' interest in the practice attaches increasing focus on humans. As actors they move on technically-oriented platforms, constituting in this environment the object of the theory-led consideration. The interest in technology use and technology application in extended contexts replaces the interest in technical principles and functioning. In changing perspectives designers and norm makers, producers and distributors, spectators and consumers, decision makers and disposers of technology are becoming actors who influence each other in communication. This may occur with different interests and values as well as subject to different conditions. In this sense, discourse communities and the changing shapes are becoming topics of research. Both interdisciplinary and intercultural approaches are demanded here.

References

Adamzik, Kirsten
2004 *Textlinguistik. Eine einführende Darstellung.* Niemeyer: Tübingen.
Agricola, Georg
[1928/1556] 2003 *De Re Metallica Libri XII. Zwölf Bücher vom Berg- und Hüttenwesen.* (Reprint of first edition VDI-Verlag 1928; Orig. 1556 latin; 1557 german) Wiesbaden: fourierverlag.
Ament, Kurt
2003 *Single Source Publishing.* New York: William Andrew Publ., Borwich.
Arntz, Reiner, Picht, Heribert and Mayer, Felix
2004 *Einführung in die Terminologiearbeit.* Hildesheim: Olms.
Media, AVIDES
2005 *MindMaps, ConceptMaps, Semantische Netze. Version 1.0* AVIDES Media AG, Rosenheim.
Bammé, Arno
2008 *Technology and Society Yearbook of the Institute for Advanced Studies on Science.* München u.a.: Profil Verlag.
Banse, Gerhard and Grunwald, Armin
2010 *Technik und Kultur. Bedingungs- und Beeinflussungsverhältnisse.* (Karlsruher Studien Technik und Kultur) Karlsruhe: Scientific Publishing.
Barnum, Carol M. and Li, Huilin
2006 Chinese and American Technical Communication: A Cross-Cultural Comparison of Differences. *Technical Communication*, Vol. 53, Number 2, 2006 May, 143–166.
Bartlett, Frederic C.
1932 *Remembering: a study in experimental and social psychology.* Cambridge: Cambridge University Press.
Bauer, Carl Otto
2000 *Rechtliche Anforderungen an Benutzerinformationen.* Lübeck: Schmidt-Römhild.
de Beaugrande, Robert-Alain and Dressler, Wolfgang
1981 *Einführung in die Textlinguistik.* Tübingen: Niemeyer.
Böhler, Klaus
2008 Die Strukturierungsmethode Information mapping (IMAP). In: J. Muthig (Hg.), 143–163.
Brandt, Edmund
2005 *Gute fachliche Praxis – zur Standardisierung von Verhalten.* Berlin: BWV Berliner Wissenschaftsverlag.
Brinker, Klaus
2005 *Linguistische Textanalyse. Eine Einführung in Grundbegriffe und Methoden.* (6. Auflage, 1. Auflage 1985) (Grundlagen der Germanistik 29.) Berlin: Schmidt.
Brosius, Hans B.
2005 *Methoden der empirischen Kommunikationsforschung. Eine Einführung.* Wiesbaden: Verlag für Sozialwissenschaften.
Budin, Gerhard
1996 *Wissensorganisation und Terminologie.* Tübingen: Narr.
Bühler, Karl
[1934] 1999 *Sprachtheorie. Die Darstellungsfunktion der Sprache.* (UTB Stuttgart 1159.) Jena: Fischer. Neudruck: 1982. Erstdruck 1934.

Burkart, Roland
2007 *Kommunikationstheorien: Ein Textbuch zur Einführung.* 4. Auflage. Wien: Braunmüller.
Carter, Ronald
2006 *Working with texts. A core introduction to language analysis.* London: Routledge.
Closs, Sissi
2007 *Single Source Publishing. Topicorientierte Strukturierung und DITA.* Siegen/ Paderborn: entwickler.press.
Dahm, Markus
2006 *Grundlagen der Mensch-Computer-Interaktion.* München: Pearson Studium.
van Dijk, Teun A.
1980 *Textwissenschaft. Eine interdisziplinäre Einführung.* München: dtv.
van Dijk, Teun A. (ed.)
1997 *Discourse as structure and process. A multidisciplinary Introduction.* London: SAGE.
DIN 2342
2004 *Begriffe der Terminologielehre.* Berlin: Beuth.
DIN EN ISO 9241
1999 *Anforderungen an die Gebrauchstauglichkeit (Bildschirmgeräte).* Berlin: Beuth.
DIN Taschenbuch 351
2008 *Technische Dokumentation. Normen für Produktdokumentation und Dokumentenmanagement.* Berlin: Beuth.
Dutke, Stephan
1994 *Mentale Modelle: Konstrukte des Wissens und Verstehens. Kognitionspsychologische Grundlagen für die Software-Ergonomie.* Göttingen: Verlag für Angewandte Psychologie.
Ehlich, Konrad
2007 *Sprache und sprachliches Handeln.* Berlin: de Gruyter.
Eigler, Gunther, Jechle, Thomas, Kolb, Monika and Winter, Alexander
1997 *Textverarbeiten und Textproduzieren. Zur Bedeutung externer Information für Textproduzieren, Text und Wissen.* Tübingen: Narr.
Faßler, Manfred
1997 *Was ist Kommunikation?* München: UTB Fink.
Felber, Helmut and Budin, Gerhard
1989 *Terminologie in Theorie und Praxis.* Tübingen: Narr.
Felix, D. and Uhr, M.
2000 Interaktion, Benutzer und Produkt im Einklang: Benutzungsfreundlichkeit im Entwicklungsprozess verankern. In: Timpe, Willumeit and Kolrep (Hg.), 317–327.
Fiehler, Reinhard (ed.)
1998 *Verständigungsprobleme und gestörte Kommunikation.* Opladen: Westdeutscher Verlag.
Frei, Viktor
2008 Was Styleguides der IT-Riesen regeln. *Sprache und Datenverarbeitung* 2: 27–51.
Geisenhanslüke, Achim and Rott, Hans (ed.)
2008 *Ignoranz: Nichtwissen, Vergessen und Missverstehen in Prozessen kultureller Transformationen.* Bielefeld: transcript.

Geiser, Georg
1998 Mensch-Maschine-Kommunikation: Warum entstehen „benutzungsfeindliche" Geräte? In: *ITG Fachbericht 154. Technik für den Menschen, Gestaltung und Einsatz benutzungsfreundlicher Produkte*, 7–20, Berlin: VDE-Verlag.
Gerson, Sharon J. and Gerson, Steven M.
2006 *Technical writing. Process and product.* Upper Saddle River, NJ: Pearson Prentice Hall.
Gerstenmaier, Jochen and Mandl, Heinz
1995 Wissenserwerb unter konstruktivistischer Perspektive. *Zeitschrift für Pädagogik* 41: 867–888.
Gerzymisch-Arbogast, Heidrun
1996 *Termini im Kontext.* Tübingen: Narr.
Göpferich, Susanne
2004 *Qualität fachsprachlicher Kommunikation.* Tübingen: Narr.
Göpferich, Susanne and Schmitt, Peter A.
1996 Begriff und adressatengerechte Benennung: Die Terminologiekomponente beim Technical Writing. In: Krings (Hg.), *Wissenschaftliche Grundlagen der Technischen Kommunikation*, 369–402. Tübingen: Narr.
Götzer, Klaus, Schneiderath, Udo, Maier, Berthold and Komke, Torsten
2004 *Dokumentenmanagement.* Heidelberg: d.punkt.
Grünwied, Gertrud
2007 *Software-Dokumentation. Grundlagen – Praxis – Lösungen.* (Kontakt & Studium 668.) 2., aktualisierte Auflage. Renningen: Expert.
Gutermann, Barbara
1996 *Die Raumdarstellung in deutschen und französischen technischen Texten.* Frankfurt: Lang.
Haase, Jana
2005 *Computer aus Nutzerperspektive. Von der Nutzeranalyse zum Interface-Design.* Wiesbaden: Deutscher Universitäts-Verlag.
Hackos, JoAnn T.
2002 *Content Management for Dynamic Web Delivery.* New York: Wiley.
Hackos, JoAnn T. and Redish, Janice C.
2000 *User and task analysis for interface design.* New York: Wiley.
Hackos, JoAnn T. and Stevens, Dawn M.
1997 *Standards for online communication. Publishing information for the Internet/World Wide Web, help systems/corporate intranets.* New York: Wiley
Hansen, Klaus P.
2011 *Kultur und Kulturwissenschaft. Eine Einführung.* 4. Auflage. Tübingen: Francke.
Hayhoe, George F.
2006 Needed Research in Global Technical Communication. *Technical Communication*, Vol. 53, Number 2, 2006 May, 141–142.
Heinecke, Andreas M.
2004 *Mensch-Computer-Interaktion.* Mit 18 Tabellen. München: Fachbuchverlag Leipzig im Carl Hanser Verlag.
Heinemann, Margot and Heinemann, Wolfgang
2002 *Grundlagen der Textlinguistik. Interaktion – Text – Dialog.* Tübingen: Niemeyer.
Hennig, Jörg and Tjarks-Sobhani, Marita (eds.)
1999 *Verständlichkeit und Nutzungsfreundlichkeit von technischer Dokumentation.* Lübeck: Schmidt-Römhild.

Hennig, Jörg and Tjarks-Sobhani, Marita (eds.)
2000 *Qualitätssicherung von technischer Dokumentation*. Lübeck: Schmidt-Römhild.
Hennig, Jörg and Tjarks-Sobhani, Marita (eds.)
2003 *Visualisierung in Technischer Dokumentation*. Lübeck: Schmidt-Römhild.
Hennig, Jörg and Tjarks-Sobhani, Marita (eds.)
2007 *Usability und Technische Dokumentation*. Lübeck: Schmidt-Römhild.
Hennig, Jörg and Tjarks-Sobhani, Marita (eds.)
2008 *Terminologiearbeit für Technische Dokumentation*. Lübeck: Schmidt-Römhild.
Herzog, Gottfried and Mühlbauer, Holger (eds.)
2007 *Normen für Übersetzer und technische Autoren*. DIN, Deutsches Institut für Normung. Berlin: Beuth.
Hoffmann, Lothar
1985 *Kommunikationsmittel Fachsprache. Eine Einführung*. Tübingen: Narr.
Hoffmann, Lothar, Kalverkämper, Hartwig and Wiegand, Herbert E. (eds.)
1998/1999 *Ein internationales Handbuch zur Fachsprachenforschung und Terminologiewissenschaft*. 2 Bde, Berlin: de Gruyter.
Hoft, Nancy
1997 *International Technical Communication*. New York: Wiley.
Hörning, Karl H.
2001 *Experten des Alltags. Die Wiederentdeckung des praktischen Wissens*. Weilerswist: Velbrück Wissenschaft.
Hubig, Christoph and Poser, Hans (eds.)
2007 *Technik und Interkulturalität. Probleme, Grundbegriffe, Lösungskriterien*. Düsseldorf: VDI (Verein deutscher Ingenieure).
Ibekwe-SanJuan, Fidelia
2007 *Application-Driven Terminology Engineering*. Amsterdam u.a.: Benjamins.
Irrgang, Bernhard
2005 *Technik im südlichen und östlichen Asien. Philosophisch-kulturelle Aspekte von Technikhöhe und Techniktransfer zwischen Europa, Indien und China*. Dresden: w.e.b.
Jacko, Julie A. (ed.)
2007 *Interaction design and usability*. Berlin: Springer.
Jahr, Silke
1993 *Das Fachwort in der kognitiven und sprachlichen Repräsentation*. (Allgemeine Literatur- und Sprachwissenschaft 1.) Essen: Die Blaue Eule.
Jahr, Silke
2007 Vom Fachwissen zum technischen Text. Prinzipien der Vertextung. In: Villiger and Gerzymisch-Arbogast (Hg.), 37–56.
Janich, Nina (ed.)
2008 *Textlinguistik. 15 Einführungen*. Narr, Tübingen.
Jansen, Carel
2002 Reflecting on Information Mapping: Does the Method Live Up to the Expectations? In: *Proceedings of the International Professional Communication Conference, 2002 (IPCC)*, 307–318.
Johnson-Laird, Philip
1983 *Mental models: towards a cognitive science of language, inference, and consciousness*. Cambridge MA: Harvard University Press.

Juhl, Dietrich
2005 *Technische Dokumentation. Praktische Anleitungen und Beispiele.* Illustriert von Tiki Küstenmacher. (VDI-Buch) 2., neu bearbeitete Auflage mit 135 Abbildungen. Berlin/Heidelberg/New York: Springer.

King, Brandy E. and Reinold, Kathy
2008 *Finding the Concept, Not Just the Word: A librarian's guide to Ontologies and semantics.* Oxford: Chandos.

Kitzmann, Andreas
2006 *Hypertext Handbook. The straight story.* New York: Lang.

Knapp, Karlfried, Antos, Gerd, Becker-Mrotzek, Michael, Deppermann, Arnulf, Göpferich, Susanne, Grabowski, Joachim, Michael, Klemm and Villiger, Claudia (eds.)
2011 *Angewandte Linguistik. Ein Lehrbuch.* Tübingen/Basel: Francke.

Köhler, Reinhard
2007 *Aspects of automatic text analysis.* Berlin/Heidelberg u.a.: Springer.

Konerding, Klaus-Peter
1993 *Frames und lexikalisches Bedeutungswissen. Untersuchungen zur linguistischen Grundlegung einer Frametheorie und zu ihrer Anwendung in der Lexikographie.* Tübingen: Niemeyer.

Krallmann, Dieter and Ziemann, Andreas
2001 *Grundkurs Kommunikationswissenschaft.* München: Fink.

Larnvik, G. M.
1996 A fairy tale on wheels. The car as a vehicle for meaning within a Norwegian subculture. In: Lie, M. and Sorensen, K.H. (eds.), *Making technology our own? Domesticating technology into everyday life*, 151–170. Oslo: Scandinavian University Press.

Lenke, Nils, Lutz, Hans-Dieter and Sprenger, Michael
1995 *Grundlagen sprachlicher Kommunikation. Mensch – Welt – Handeln – Sprache – Computer.* München: Fink.

Lum, Casey Man Kong
2006 *Perspectives on Culture, Technology and Communication. The Media Ecology Tradition.* Creskill, NJ: Hampton Press.

Lundquist, Lita and Jarvella, Robert J. (eds.)
2000 *Language, Text, and Knowledge.* Berlin: Mouton de Gruyter.

Mansour, Markus
2006 *Informations- und Wissensbereitstellung für die lebenszyklusorientierte Produktentwicklung.* Essen: Vulkan.

McCool, Matthew
2006 Information Architecture: Intercultural Human Factors. *Technical Communication*, Vol. 53, Number 2, 2006 May, 167–183.

Metzing, Dieter (ed.)
1980 *Frame conceptions and text understanding.* Berlin: de Gryuter.

Miller, George A.
1956 The magical number seven, plus or minus two: Some limits in our capacity for processing information. *Psychological Review* 63/2, 81–97.

Min-Sun, Kim
2002 *Non-western perspectives on human communication. Implications for theory and practice.* Thousand Oaks, Calif.: Sage Publ.

Misoch, Sabina
2006 *Online-Kommunikation*. Konstanz: UTB.
Muthig, Jürgen (ed.)
2008 *Standardisierungsmethoden für die Technische Dokumentation*. Lübeck: Schmidt-Römhild.
Muthig, Jürgen and Schäflein-Armbruster, Robert
2008 Funktionsdesign – methodische Entwicklung von Standrads. In: Muthig (Hg.), 41–73.
Nielsen, Jakob
1996 *Multimedia, Hypertext und Internet – Grundlagen und Praxis des elektronischen Publizierens*. Braunschweig: Viehweg.
Picht, Heribert (ed.)
2006 *Modern Approaches to Terminological Theories und Applications*. Frankfurt: Lang.
Pogner, Karl-Heinz
2012 The social perspective in the production of technical text and knowledge: disourse communities and communities of practice. In: A. Rothkegel and Ruda, S. (eds.), *Communication on and via technology*. Berlin: Mouton de Gruyter.
Pörings, Ralf and Schmitz, Ulrich (eds.)
2003 *Sprache und Sprachwissenschaft. Eine kognitiv orientierte Einführung*. Tübingen: Narr.
Poncini, G.
2004 *Discursive strategies in multicultural business meetings*. Bern: Lang.
Rehbein, Jochen
1977 *Komplexes Handeln. Elemente zur Handlungstheorie der Sprache*. Stuttgart: Metzler.
Reinicke, Tamara
2004 *Möglichkeiten und Grenzen der Nutzerintegration in der Produktentwicklung. Eine Systematik zur Anpassung von Methoden zur Nutzerintegration*. München: Verl. Dr. Hut.
Rickheit, Gert (ed.)
1991 *Kohärenzprozesse. Modellierung von Sprachverarbeitung in Texten und Diskursen*. Opladen: Westdeutscher Verlag.
Roelcke, Thorsten
2005 *Fachsprachen*. Berlin: Erich Schmidt.
Rösch, Olga (ed.)
2008 *Technik und Kultur*. Berlin: News & Media.
Rogge, H.
1980 „Ein Motor muss aussehen wie ein Geburtstagsgeschenk." In: T. Buddensieg (Hg.), *Industriekultur: Peter Behrens und die AEG*, 91.126. Bonn: Gebr. Mann Verlag.
Rothfuss, Gunther
2002 *Content Management mit XML. Grundlagen und Anwendungen*. Berlin: Springer.
Rothkegel, Annely
1993 *Object Knowledge and Text Knowledge*. London: Pinter.
Rothkegel, Annely
2007 Kulturspezifische Technikkommunikation. In: Straub, Jürgen, Weidemann, Arne, Arne and Weidemann, Doris (Hg.), *Handbuch interkulturelle Kommunikation und*

Kompetenz. Grundbegriffe – Theorien – Anwendungsfelder, 604–613. Stuttgart: Metzler.

Rothkegel, Annely
2008 Wissenssysteme und ihre konzeptuelle Transformationen in der Experten-Nichtexpertenkommunikation: Technikkommunikation in kultureller Perspektive. In: Rösch (Hg.), 48–60.

Rothkegel, Annely
2010 *Technikkommunikation. Produkte, Texte, Bilder*. (UTB 3214.) Konstanz: Huter & Roth.

Schmidt, Siegfried, J. and Zurstiege, Guido
2007 *Kommunikationswissenschaft: Systematik und Ziele*. Hamburg: Reinbeck.

Schmidt, Vasco
2012 Software documentation: the role of computers as a means, tool and medium. In: A. Rothkegel and S. Ruda (eds.), *Communication on and via technology*. Berlin: Mouton de Gruyter.

Schmitt, Peter A.
1995 Warnhinweise in deutschen und englischen Anleitungen: Ein interkultureller Vergleich. *Fremdsprachen Lehren und Lernen* 24, 197–222.

Schmitt, Peter A.
1999 *Translation und Technik*. Tübingen: Stauffenberg.

Schnotz, Wolfgang
1994 *Aufbau von Wissensstrukturen. Untersuchungen zur Kohärenzbildung bei Wissenserwerb mit Texten*. Weinheim: Beltz.

Schröder, Thomas
2003 *Die Handlungsstruktur von Texten. Ein integrativer Beitrag zur Texttheorie*. Tübingen: Narr.

Schubert, Klaus
2007 *Wissen, Sprache, Medium, Arbeit. Ein integratives Modell der ein- und mehrsprachigen Fachkommunikation*. Tübingen: Narr.

Schulz von Thun, Friedemann
2004 *Miteinander reden. Kommunikationspsychologie für Führungskräfte*. Boniswil: Rusch.

Schwender, Clemens
2003 Zur Geschichte von Abbildungen in Instruktionstexten. In: Hennig and Tjarks-Sobhani (Hg.), 32–58.

Schwender, Clemens
2012 Instructional illustrations: Waypoints in the historical development. In: A. Rothkegel and Ruda, S. (eds.), *Communication on and via technology*. Berlin: Mouton de Gruyter.

Searle, John R.
1969 *Speech Acts – An Essay in the Philosophy of Language*. Cambridge: Cambridge University Press.

Shannon, Claude and Weaver, Warren
[1949] 1995 *The mathematical theory of communication*. Urbana: University of Illinois Press.

Staffeldt, Sven
2008 *Einführung in die Sprechakttheorie. Ein Leitfaden für den akademischen Unterricht*. Tübingen: Stauffenburg.

Stein, Stephan
2003 Textgliederung, Einheitenbildung im geschriebenen und gesprochenen Deutsch: Theorie und Empirie. (*Studia Linguistica Germanica* 69.) Berlin: de Gruyter.
Stenschke, Oliver and Wichter, Sigurd (ed.)
2009 *Wissenstransfer und Diskurs*. Frankfurt: Lang.
Stieniczka, Norbert
2006 *Das "narrensichere" Auto. Die Entwicklung passiver Sicherheitstechnik in der Bundesrepublik Deutschland*. Darmstadt: Wissenschaftliche Buchgesellschaft.
Timpe, Klaus-Peter, Willumeit, Hans-Peter and Kolrep, Harald (ed.)
2000 *Bewertung von Mensch-Maschine-Systemen*. Düsseldorf: VDI-Verlag.
Ulijn, Jan M. and Strother, Judith B.
1995 *Communicating in technology and business: From psycholinguistic theory to international practice*. Bern: Lang.
Villiger, Claudia
2007 Von der Kaffeemaschine zum Textverarbeitungsprogramm: Risikokommunikation in Gebrauchsanleitungen und Online-Hilfesystemen. In: Villiger and Gerzymisch-Arbogast (Hg.), 129–149.
Villiger, Claudia
2008 Schlüsselwörter zum Wissen: Mehrsprachige Terminologien kooperativ erstellen und nutzen. *Sprache und Datenverarbeitung* 2 : 9–26.
Villiger, Claudia and Gerzymisch-Arbogast, Heidrun (ed.)
2007 *Kommunikation in Bewegung. Multimedialer und multilingualer Wissenstransfer in der Experten-Laien-Kommunikation*. Frankfurt: Lang.
Wagner, Jörg
2002 *Mensch-Maschine-Interaktion. Sprachwissenschaftliche Aspekte*. Frankfurt: Lang.
Weigand, Edda
2004 *Sprache als Dialog. Sprechakttaxonomie und kommunikative Grammatik*. Tübingen: Niemeyer.
Weissgerber, Monika
2010 *Schreiben in technischen Berufen*. Erlangen: Publicis.
Wendt, Susanne
1997 *Terminus – Thesaurus – Text. Theorie und Praxis von Fachbegriffssystemen und ihrer Repräsentation in Fachtexten*. Tübingen: Narr.
Wright, Georg Henrik von
[1968] 1977 *Handlung, Norm und Intention. Untersuchungen zur deontischen Logik*. Berlin: de Gruyter. (orig. engl. 1968)
Wright, Sue Ellen and Budin, Gerhard (ed.)
1997/2001 *Handbook of Terminology Management*. Vol. I/II. Amsterdam: Benjamins.
Ziem, Alexander
2008 *Sprachliches Wissen und Frames. Kognitive Aspekte der semantischen Kompetenz*. Berlin: de Gruyter.

Franck Ganier

Cognitive models of processing procedural instructions

1 Introduction

During the last thirty years or so, several studies have been published in Cognitive Psychology, Psycholinguistics and Ergonomics on the cognitive processes involved in processing procedural instructions (including computer documentation, online-help, assembly tasks).

These studies have led to a better understanding of these kinds of processes, differentiating them from those used with other kinds of text such as narrative (novels, reports) or didactic texts (textbooks, encyclopaedias). Some models, based on these studies, have been developed in detail in order to explain how users process the information conveyed in instructions – from their initial reading to carrying out the prescribed actions.

The aim of this chapter is to present a comprehensive view of the main models proposed by researchers to describe the mental activities implicated in the use of procedural documents, or more precisely, in the user/document/equipment interaction. After first describing the methods employed for studying the use and comprehension of procedural instructions, six processing models are presented, ranging from local and serial to global and then to interactive ones. In conclusion, I suggest that these models could be used within three different perspectives: (i) basic or applied research, in order to enhance our knowledge of cognitive processes and factors affecting them; (ii) the training of technical writers, in order to improve the design of procedural instructions; and (iii) the training of future readers, in order to facilitate their use.

Performing a task with the help of a procedural document is a common daily activity. Using a piece of software, programing a VCR, assembling a DIY kit, or cooking a meal may seem to be simple activities, but they actually hide many complex cognitive processes. Indeed, when an individual reads a set of instructions and then carries them out, he or she has to transform the linguistic, graphic or audiovisual information into an action or a set of actions in order to perform the task correctly. This transformation is not always obvious since the use (or the non use!) of procedural documents often leads to small or large and important failures. The reason for these failures is often explained by the fact that the documents do not fit the needs or the cognitive processes of the user, despite the fact that technical writers usually feel that they have correctly applied known principles of technical writing (Ganier and Barcenilla 2007).

Studying how people use procedural instructions is important because of their finality, i.e., users can perform successfully the described task. Up until the end of the 1970ies, the research in this domain was often carried out from an ergonomic perspective (for example, see Sremec 1972). Researchers were more concerned with studying general performance related to the variations in instructional formats than on studying the cognitive processes involved in their use. In the beginning of the 1980ies, however, some researchers, following Wright and Wilcox (1978), began to consider that, in order to improve procedural documents, it was necessary to take into account the way users used these documents and processed the information they contained. Other researchers (e.g., Kieras and Bovair 1986 and Dixon 1987) showed more interest in studying the cognitive processes required to process this kind of text from a basic research perspective.

Thus, from the 1970ies, a considerable amount of research has been done in Cognitive Psychology focusing on reading and applying procedural instructions. This research has allowed us to develop some general models aimed at describing the processes and interactive mechanisms that occur between the encoding of the information and the execution of the appropriate corresponding actions. The identification of these processes or mechanisms has a double advantage. On the one hand, it allows us to increase our knowledge in this domain and to use this knowledge in an educational perspective. On the other hand, it allows us to identify the sources of problems, which when identified, can lead to design recommendations for professional technical writers. Keeping this double perspective in mind, the goal of this chapter is to present the main models proposed by selected researchers to describe the mental activities implied in the use of procedural documents, or more precisely, in the interaction between the user, the procedural document, and the equipment involved.

2 Methods for studying the use and understanding of instructions

Studying the processing of instructions can be done at two levels. The global level essentially uses observations and analyses of the (behavioral) activity of the user in trying to study in a general way the behaviors implicated in following instructions. The local level is more focused on the processing of instructions in more rigorously controlled experimental settings. The approach at this level aims to study the cognitive processes implied in the reading and the understanding of the instructions, and in the execution of corresponding actions.

At the global level, a commonly used technique to observe the use of instructions involves video recording of individuals performing a given task and analyzing their activities on the basis of collected behavioral protocols. This approach can focus on the number, and the time spent on inspecting the document and the equipment; on the number, and the time spent on carrying out the actions; on the arrangement of these different kinds of activities and on the procedures followed by the user compared to the expected ones. These data can be collected in more or less controlled situations according to the goals and hypotheses of the researcher. In the most ecological situations (very close to reality), the user is asked to perform a set task without being given any information about how to follow the instructions. In the most controlled situations (very close to experimental settings), the user can be asked to use a booklet that contains instructions (e.g., with each page presenting a new task to carry out), and to follow them carefully (e.g., see Ganier, Gombert and Fayol 2000). In this situation, the researcher does not give the participant any information about the task to perform. He or she has to discover this from following the instructions.

At the local level, in studies focusing more specifically on the cognitive processes implicated in the reading and comprehension of instructions, behavior related to the reading or to the execution of the instructions can be collected in real time (on-line methods) or with a delay (off-line methods). On-line studying of the reading and comprehension of procedural instructions aim to analyze reading, comprehension and execution processes at the moment they take place. Different techniques are generally used. Thus, video recordings, mentioned above, have often been used in experimental settings. However, to overcome any lack of precision, new experimental devices have been designed to collect and analyze more precise behavioral data. Thus, in the 1980ies, self-paced presentation techniques were developed. These techniques allowed the user to see the next part (or "segment") of a text containing some instructions on a computer screen by pressing a key on the keyboard (cf. Dixon 1987). The computer collected the display times for each segment whilst the participants performed the task in the natural environment (for example, using a paper-pencil drawing task or operating an actual piece of equipment). These methods evolved into screen presentations of the instructions accompanied by screen simulations of the devices being manipulated (Duggan and Payne 2001). In some cases, the instructions can be blurred and a click on them makes them more readable whilst measuring their display times (see the "click and read" method proposed by Ummelen 1997). In other cases, the screen displays a working zone (i.e., a screen window where the user can carry out his/her actions) that masks the window displaying the instructions. Then the computer alternately computes the display times for the instructions and for the actions (Ganier 2000/2001). Eye-movement recording, used for a long time to study the reading of narrative or expository texts, also began to be applied about this time to the domain of reading and comprehension of procedural instructions (Schmid and Baccino 2001; Alamargot et al. 2006). This method consists of collecting gaze durations and locations as well as movements of the user's eyes while consulting the instructions.

Whatever the technique, the principle remains the same: i.e., we determine the time the user needs to process (to read and to understand) a part or all of the instructions that are presented under various conditions. The collected behavioral data are very precise and allow to determine the nature, the chronology and the length of the processes involved. Complementary to the behavior implied in the reading of the instructions, the actions performed by the users also provide interesting information about their cognitive processes. For example, because these actions immediately follow the reading of the instructions, they inform the researcher about the different states of representation that the users elaborate from the instructions and, therefore, they allow a step-by-step tracking of the processes of understanding.

The off-line studying of reading and understanding procedural instructions consists of collecting some data after the reading episode has taken place (immediately after or with a varying delay). For example, various measures of the user's activity can be collected such as task success or failure, or the number of errors made. The user can also be asked to give a verbal or a motor recall of what was read or carried out. These techniques specifically allow the researchers to obtain some information about the organization of procedures in memory.

Research on the use of instructions, and on their reading and understanding, helps us to identify some of the cognitive processes implicated in their processing. Historically, early researchers were interested in very local phenomena bound to the reading and comprehension of instructions, and they proposed a very linear and sequential modeling of these processes. Essentially this involved following a kind of schema beginning with the reading of the instructions and ending with the execution of the corresponding actions. As the research progressed, however, the models became more global, presenting more interactive processes, and not necessarily beginning with a reading episode. In the following parts of this chapter, this evolving perspective will be used to describe the development of cognitive models of the processing of procedural instructions.

3 Serial and local models: From reading to execution

Three models are presented in this part: these are those of Wright and Wilcox (1978), Kieras and Bovair (1986) and Dixon, Harrison and Taylor (1993). The first two models focus on the local processing of procedural instructions and are called serial models. This is because they consider that processing takes place in a strict and determined order. According to these models, processing starts with reading an instruction and ends with executing the corresponding action. (In these models, this last stage is often the "application of the knowledge gained from reading", suggesting the linear and sequential approach). The third model also focuses on the local processing of instructions but, in this case, the processing does not operate in a strictly serial way from the initial reading to the final application of the knowledge gained. It moves toward a more interactive model where both the user's knowledge and information coming from the user's environment (for example, from the equipment) play a role in the elaboration of mental representations underlying the intended action.

3.1 A first framework: The Wright and Wilcox (1978) model

Wright and Wilcox (1978) published one of the first studies that focused on the local processing of procedural instructions by attempting to identify the main cognitive processes implied in reading and carrying out written instructions. The goal of these researchers was to identify different time periods occurring in the overall process of following procedural instructions, and to determine which kinds of cognitive processes might be dominant during these periods. In one study, they observed that when adults were asked to carry out a simple written instruction such as "Draw a circle above a square", the respondents regularly interrupted their drawing with reading activities and pauses. Wright and Wilcox thus proposed three distinct time periods to analyze the reader's behavior: the first corresponding to the time spent reading the instruction (labeled "reading period"), the second corresponding to the time spent thinking about the instruction (labeled "thinking period"), and the third corresponding to the time spent carrying out the action (labeled "drawing period" for this specific task). According to Wright and Wilcox these three distinct time periods roughly cover three kinds of cognitive processes:

- processes occurring in the reading of the instructions;
- processes occurring in integrating the information and in action planning;
- processes occurring in carrying out the action.

The framework described by Wright and Wilcox (1978) was supported by observations and experiments conducted by other researchers. For example, Vermersch (1985) analyzed how adults followed an unfamiliar recipe. Vermersch observed that the participants did not memorize straightaway the whole set of instructions necessary to carry out the recipe in one go, but encoded one step at a time from the text, carried it out, checked it, and then encoded the following step, and so forth. Vermersch called this phenomenon – switching between the instructions and carrying out the task, leading to a systematic fragmentation of the execution phase in elementary actions – the "atomization of the action". For some researchers, this iterative process was considered as a strategy or a necessary adaptation aimed at reducing the cognitive demands on working memory (e.g., Guthrie, Bennett and Weber 1991; Boekelder and Steehouder 1999; Duggan and Payne 2001; Ganier 2004).

3.2 "Translating" the instructions into actions: The Kieras and Bovair (1986) model

In the middle of the 1980ies, Kieras and Bovair (1986; Bovair and Kieras 1991) proposed a model centered on the stage of "translating" the instructions being read in a representation of the described procedure. This model specifies the cognitive processes identified by Wright and Wilcox (1978). It presented a cognitive architecture (based on Anderson's ACT structure, 1983) in which text processing processes would take place, similar to those proposed by Kintsch and van Dijk (1978; van Dijk and Kintsch, 1983) and Just and Carpenter (1987).

3.2.1 The architecture of the system: Knowledge and memory registers

Kieras and Bovair's (1986) model (see Figure 1), based on the theoretical propositions of Anderson (1983), distinguishes between two kinds of knowledge (declarative and procedural), and three memory registers: declarative, procedural and working memory. Declarative knowledge, defined as "knowing that", is assumed to be stored as a semantic – or propositional – network in

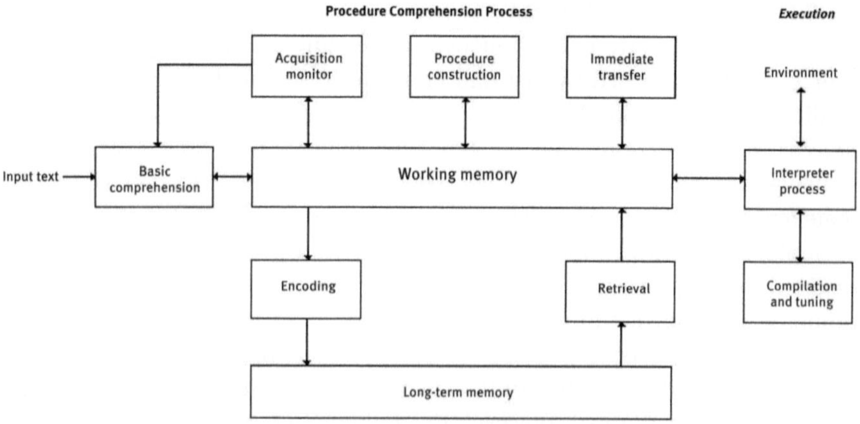

Figure 1: Kieras and Bovair's (1986) model of acquiring procedures from text (from Bovair and Kieras, 1991: 210)

a long term memory register named "declarative memory". Procedural knowledge, defined as "know-how", is assumed to be stored in the form of "production rules" in another long term memory register – named "procedural memory". "Working memory" (Baddeley 1986) is related to these two memory registers. Working memory provides a temporary storage and workspace allowing for real time processing of the information coming in from the environment and the individual's long term memory registers. When carrying out procedural tasks, working memory is dedicated to processing the information coming in from the declarative and procedural memory registers, from encoding the situation elements, and from applying the production rules, as well as from the awaited results or the user's goals (Engle, Carullo and Collins 1991; Glover, Harvey and Corkill 1988).

3.2.2 Cognitive processes implied in the acquisition of procedures from text

Kieras and Bovair (1986) distinguish between two main comprehension processes: a basic reading comprehension process and a procedure comprehension process. They consider that the former produces a propositional representation of the input text (Kintsch and van Dijk 1978; van Dijk and Kintsch 1983). This propositional representation is then used by the latter to elaborate a representation of the procedure that can be executed by the so called "interpreter". Once the procedure can be correctly executed, knowledge compilation process can occur (Anderson 1983).

3.2.2.1 Reading comprehension

In the Kieras and Bovair (1986) model, the reading comprehension process is similar to that proposed by Just and Carpenter (1987) and it can apply to the reading of different kinds of texts (expository, narrative). Readers read and process the instructions step by step, one sentence at a time, parsing each sentence, and doing the basic referential and semantic analysis needed to elaborate a propositional representation for each sentence in working memory. For example, a referential analysis can consist of relating the label of a particular object to an appropriate concept. Thus, a knob could be referred to as the tuning knob, and this label must be attached to an instance of the concept KNOB. Moreover, in order to carry out the procedure, the actual physical object referred to must be identified in the environment, so that the specified action can be actually performed. However, the authors do not specify if this identification occurs during the comprehension process or later.

3.2.2.2 Procedure comprehension

In the Kieras and Bovair (1986) model, procedure comprehension processes elaborate a declarative representation of the procedure from the propositional representation of the text. Procedure comprehension consists of three major sub-processes. The procedure construction process takes the representation of the text and elaborates the declarative form of production rules, that is an executable propositional representation of the procedure. The immediate transfer process checks if newly constructed rules are already known. It compares the representation of the current rule to the already-known rules. If the rule is new, then it must be maintained in working memory until it is stored in long-term memory, which is time consuming. If the current rule is very similar to an existing rule (or if it already exists), then some (minor) modifications of the existing rule will be required, but this will take a shorter time (Johnson and Kieras 1983). This immediate transfer process is responsible for large time savings when learning new procedures if they have steps in common with previously read or learned procedures. The third sub-process is the acquisition monitor process, which monitors whether a new rule has been fully learned. This process monitors (a) the acquisition of the steps of the procedure, (b) the distinction between acquired and not yet acquired steps and (c) the achievement of the acquisition of a new rule.

3.2.2.3 Procedure execution and debugging

Once the declarative version of a procedure has been constructed, a procedure interpretation process executes the procedure from the elaborated representation. This execution will succeed if the declarative representation has been correctly constructed. After this stage has been reached, the process of skill acquisition that creates a procedural representation of the procedure can start. This stage, supposed to result from practice and experience rather than the reading of the text, is not well specified by Kieras and Bovair (1986), who refer the reader back to the Anderson's (1983) model to account for it.

3.2.3 Contributions and limits of this model

Kieras and Bovair's (1986) model focuses on the acquisition of procedures from written instructions with novice users, and it emphasizes processes that translate basic reading comprehension into the correct comprehension of procedures. Compared to Wright and Wilcox's (1978) model, this model attempts to account for both prior knowledge and repetition of situation effects in the acquisition of procedures. Thus, two kinds of processes are added: the immediate transfer process and the acquisition monitor. Furthermore, the authors suggest some processing related to the environment (the equipment), but without specifying when and how this is taken into account by the user.

Even if this model is more detailed than that of Wright and Wilcox (1978), it is still a linear processing model of understanding and following instructions, since processing begins with reading and goes on until the actions are carried out. Indeed, Kieras and Bovair (1986; Bovair and Kieras 1991) consider that in order to carry out a procedure, the user first constructs a declarative version of the production rules. When a complete and correct declarative representation of the procedure has been constructed, the procedural representation can then be elaborated. The process of procedure comprehension consists, therefore, of generating a correct set of rules from the propositions delivered by the text. To be able to represent a procedure correctly, production rules must have a set of conditions and actions coordinated in a suitable way. For Kieras and Bovair (1986), it is necessary that the syntax of the instructions corresponds to an action plan that can be elaborated by the reader. For example: "If the meter is set to 20, turn the knob" corresponds to the production rule: "IF (condition) THEN (action)". However, when written in natural language, instructions do not always respect this syntax. Furthermore, other instructional formats for conveying procedures are possible, especially formats including pictures. However, Kieras and Bovair's model does not take in account these surface aspects that could be important in the processing of procedural documents and instructions. For example, differences in performance due to the presentation of different instructional formats for the same procedure (e.g., text vs. picture) cannot be addressed in this framework (Morrell and Park 1993; Ganier, Gombert and Fayol 2000).

3.3 A building multiple representations framework: The Dixon, Harrison and Taylor (1993) model

In the beginning of the 1990ies, Dixon, Harrison and Taylor (1993) adopted a multiple representations framework assuming that at least three kinds of representations were involved in understanding procedural texts. According to this conception, the translation process from reading to the execution of action(s) can be described as a process of construction of three levels of representations: a word-level representation, a semantic representation and a referential representation.

3.3.1 Building and holding a word-level representation

The first level of representation, described as a word-level representation, is built from reading or listening to the initial instruction and then storing it in working memory for a very brief period (a few seconds). This kind of representation preserves the formal features of the statement (lexical and syntactic features such as word order, passive or active voice, etc.). It is usually considered as a verbatim or surface-level representation because it allows the user to keep in mind the exact form of the part of the instructional text he or she has just read or heard. For example, if the instruction "Put it in reverse" is given to a car driver, immediately followed by two more statements "You must put it in reverse" and "Put it in reverse", the reader will not have any difficulty recognizing that the last statement corresponds to the initial one. On the other hand, after several minutes' delay, he or she will be unable to determine which statement was presented earlier (Ganier and Heurley 2003).

3.3.2 Building a semantic representation of the instructions

The second level of representation is the semantic representation of the information conveyed by the text. A meaning is extracted from the statement, but the exact wording is not yet preserved. In the previous example, after several minutes delay, the reader who was presented with the "Put it in reverse" statement will be able to remember that the action involved in putting the car into reverse must be achieved, but he or she will probably be unable to remember the exact wording of the instructions used to communicate this information ("You must put it in reverse", "Reverse must be engaged", "Put it in reverse").

3.3.3 Building a referential representation of the instructions

The third level of representation, defined as the referential representation of the instructions, is no longer a semantic representation of the instructions, but a representation of what the instructions refer to, that means the procedures required to carry it out. In the case of procedural texts, Dixon, Harrison and Taylor merge this third level with the action plan that is supposed to allow the execution of the actions described in the instructions. These different informational resources – the instructions provided, the environment, and the user's previous knowledge – are supposed to be used jointly, in a dynamic process that depends on the user's actions on his/her environment. Thus, understanding the statement "Put it in reverse", is to be able to represent the succession of actions needed to carry this out, and the events and states necessary to engage the reverse gear in a specific car (e.g., depress the gear lever, bring it well to the left, and then push forward, or lift the gear lever, push it to the right and then pull it right back).

3.3.4 Contributions and limits of this model

Dixon, Harrison and Taylor's (1993) model addresses the elaboration of different levels of representation (from the word-level to the referential representation of the text), taking into account the user's prior knowledge about the instructional task (e.g., the equipment to be manipulated). Compared to the two previous models, this model is no longer a strictly linear or sequential one (starting with the reading of the instructions and finishing with the application of the acquired knowledge) since both the environment and the user's knowledge are implicated in the construction of the referential representation.

4 Towards local and then global interactive models

The above models are considered as local processing models as they generally describe a linear and sequential processing of the instructions prior to executing the actions. They usually follow the pattern "read then use" but they cannot easily give an account of other processing patterns. Two more models are presented in this part: those of Guthrie, Bennett and Weber (1991) and Wright (1983, 1999a, 1999b). The former focuses on the local processing of instructions but, compared to the previous models, is more interactive since it clearly takes in account the interactions between the user, the document and the equipment. The latter, which can also be labeled interactive, adopts a more global vision of the use of procedural documents since it attempts to give an account of the processes occurring before the reading of the instructions (i.e., searching for information in the document).

4.1 The user/document/device interaction: Guthrie, Bennett and Weber's (1991) matrix model

Guthrie, Bennett and Weber (1991) described the processing of procedural instructions with the help of a matrix (see Figure 2), integrating the user's

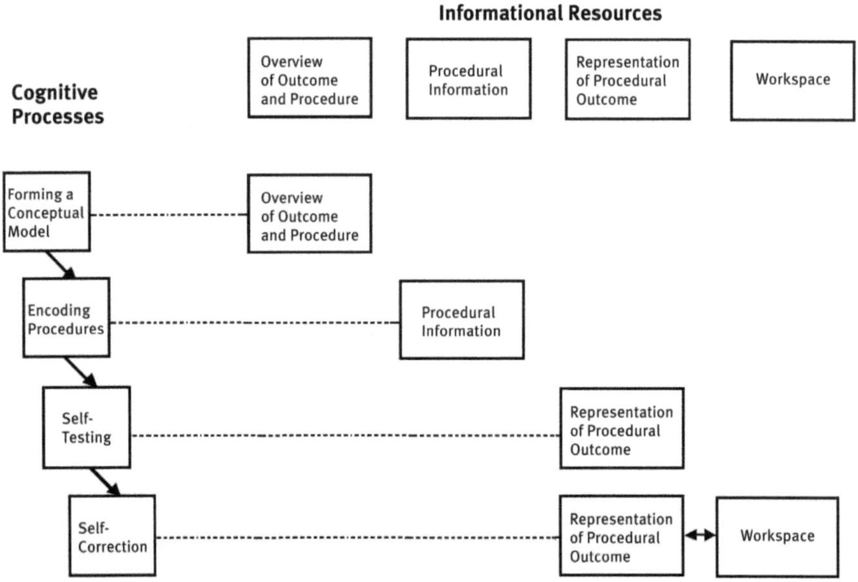

Figure 2: Guthrie, Bennett and Weber's (1991) model for following instructions

cognitive processes with the following four informational resources: (a) a general overview of the outcomes and the procedures; (b) the procedural information; (c) a representation of the result; and (d) the workspace.

4.1.1 Forming a conceptual model of the device or of the task to be achieved

According to these authors, forming a conceptual model of the device or of the task to be achieved consists of constructing a representation of the object or the issue of the task. This process does not occur automatically during the step-by-step reading of a set of written instructions nor while inspecting a diagram of a specific part of the outcome. On the contrary, this conceptual model can be developed more effectively by reading the exposition in a prose format or while examining a picture of the general outcome. Thus, for example, in a user's guide to operating a camera, elaborating the conceptual model of the equipment is afforded by the presentation of a general view of the camera.

4.1.2 Encoding procedural information

The encoding of the procedural information consists of identifying executable steps and then entering them into working memory. The authors suggest that neither a verbal nor a graphic instructional format is ideal, but that the process should be sensitive to the nature of the task. Thus, when the outcome involves verbal materials, the most efficient means to encode the procedural information should be a set of written instructions. Conversely, if the task consists of assembling a particular object, the most efficient source of procedural information should be a set of pictures. When the task involves a chain of procedures (such as driving, cooking, or solving a series of problems) then the optimal informational format is to present prominent temporal features. In cases such as these, the temporal information is frequently represented as a set of well-ordered steps (either in text or pictures). In short, according to Guthrie, Bennett and Weber, the optimal format for presenting procedural information is the one that is most compatible with the behavior needed to perform the task. Thus, the presentation format has to be defined in terms of ease in transferring information from the instructional document to the performance situation. It must aim to reduce the number and the complexity of the transformations that the cognitive system has to perform in order to complete the task. An optimal procedural document thus minimizes the transformations required of the user. Good mapping between the information presented in the document and the task to perform should facilitate the processing of the instructions. Conversely, an inadequate presentation format will impose a heavier cognitive load on the user and result in an increase in time and/or errors in performance.

4.1.3 Self-testing and feedback

Self-testing requires a comparison between every step of the procedure or the outcome to a "standard" that can be internal (for example, in memory). If an internal standard does not exist, self-testing will consist in comparing the task progression to a representation (of a part) of the outcome (if it exists). Self-testing is important in the processing of procedural documents because of the greatly constrained nature of the task. For example, in many cases, few alternatives exist for assembling a piece of equipment. Every step must be correct so that the task can be successfully performed. Guthrie, Bennett and Weber (1991) propose that, in most cases, self-testing will be optimal with a graphic representation of the issue. This combination of cognitive processes and informational resources facilitates self-testing because a picture, in contrast to text, presents spatial, locational and relational information concerning the task. So, self-testing can occur while comparing the graphic representation of the issue with the task progression in the workspace. In some cases, the steps of the procedure can be used for self-testing. Indeed, numerous steps (either verbal or graphic) contain information that allows the user to engage in self-testing in order to insure that the operations have been correctly performed.

The authors propose that, during the processing of a procedural document, it is important to engage in self-testing and gain internal feedback. For procedural documents, the authors suggest that monitoring activities must be more frequent, more precise, and better timed than the self-monitoring of comprehension that occurs with other kinds of text. However, the frequency and the different kinds of self-testing in the processing of procedural documents have not been studied directly. Moreover, the equipment has to provide information that can be used in combination with the instructions so that an internal feedback loop can help the user to perform the task.

4.1.4 Self-correction

In Guthrie, Bennett and Weber's model, self-correction consists of correcting mistakes that occur during task performance. The process requires three kinds of information: (a) a representation of the task to perform (for example, a picture of the outcome); (b) a series of instructions in a text format; and (c) the outcome. Depending on the nature of the task, self-correction can rely on the written instructions, on the pictures (that can be compared to the real features of the equipment), or on a guide aimed to help repair mistakes. However, in spite of their importance, self-correction behavior and the expertise required to repair mistakes are not precisely examined in this model.

4.1.5 Contributions and limits of this model

Guthrie, Bennett and Weber (1991) suggest that translating procedural information in a document into action can be represented by a set of interactions between the document and the cognitive processes. These operations are achieved more efficiently when the informational resources minimize the number and the complexity of the transformations required to process the information given in the document. The authors expect that different informational resources will be optimal for each of the four processing units described in their model. Thus, a general presentation of the task is likely to be effective because it facilitates the elaboration of a conceptual model. The procedural information (either verbal or graphic) that emphasizes the temporal (or sequential) features of the performance facilitates the encoding of the procedure. A representation of the outcome (in graphic format) would improve self-testing, and the comparison of the outcome with task progression would lead to more efficient self-correction processes.

4.2 Cognitive abilities implied in the use of procedural documents: Wright's model (1999a, b)

Wright and Wilcox's (1978) model tried to describe the processes implied in the reading of instructions and carrying out the corresponding actions. In 1983, Wright wished to develop this simple framework, and she outlined the bases of a model aimed at describing how adults use procedural documents in a more general way. Indeed, according to Wright, the use of procedural documents is not restricted to the reading, understanding, planning and execution of actions. Use begins before reading when the user tries to locate information in the document in order to answer to a problem that has arisen during the interaction with the equipment. Wright's model subdivides users' activities in three clusters: research, understanding, and application. For the users, or 'readers', the first need is to search for desired information: they must find particular information specific to their needs. When this information has been found, it has to be understood before it can be applied (leading to the execution of actions).

The model, developed in 1999, specifies the cognitive processes implicated in the three clusters of activities proposed earlier in 1983. Wright (1999b) underlines that this subdivision does not imply that the processes take place in a strictly linear way, but can be recycled, as shown in Figure 3.

4.2.1 Searching

Wright (1999a) assumes that the reading of procedural documents is not systematic: it is done whenever an individual tries to achieve some goal such as operating some piece of equipment or accessing a service. This reading, or more precisely this use, does not take place in a linear way from the first to the last page but is highly selective. Rather than reading exhaustively, the user's first aim is to reach specific parts of the information in the document. Therefore, using procedural documents requires the user to formulate a ques-

A. Search skills

1. Formulate search target
2. Comprehend document structure
3. Modify search target to fit document structure
4. Locate information thought relevant to target
5. Decide if further search is necessary (if so recycle from A. 1. or go to cluster B.)

B. Comprehension skills

1. Interpret sentences, assigning meaning to words
2. Infer exophoric implications
3. Interpret adjuncts to current text
4. Integrate information drom different locations, within and outside text
5. Decide if more information is needed (if so recycle from A. 1. or go to 6.)
6. Take decision about actions (if intend to act, go to cluster C.)

C. Execution skills

1. Formulate action plan
2. Segment and remember text relating to 1st part of action plan
3. Carry out 1st part of plan
4. Monitor progress as alternate between text+action until plan is completed
5. Check if desired goal is achieved (if not re-read instructions B.1.).

Figure 3: Synopsis of skills involved in the use of procedural instructions (Wright, 1999a)

tion first or, more precisely, a research target corresponding to the user's goal. When he or she has decided what must be searched for, the user has to understand the structure of the document in order to find it. Sometimes, the research target has to be modified so that it corresponds to the content and structure of the document (for example, when the research target does not match the terminology used in the index). When the information matching the search target has been found, the user considers that the search is over and moves to another task. If the information cannot be found he or she might decide that further searching may be needed (for example, after refining the research question).

4.2.2 Understanding

The stage of "reading to understand" occurs when the user has located the information that seems to offer an answer to his/her question (or seems to fit the search target). By now, the dominant cognitive processes are those related to understanding text (or pictures). Here, the user has to interpret the written instructions by assigning meaning to the words read, doing syntactic calculations, etc. The extraction of the necessary information and the integration of disparate elements of the information (that could be inside or outside the text) are crucial to understanding correctly what has been read. Here, not only cognitive, but also metacognitive skills are necessary. For example, the user needs to monitor that the located information is applicable to reach his/her goal as well as monitoring that his/her understanding of the text is correct. This is necessary in order to decide if he or she needs additional information to reach his/her goal. If this is this case, then he or she can make a fresh search for additional information. Then, decisions concerning the execution of actions can take place.

4.2.3 Applying the acquired knowledge

The user creates an action plan on the basis of what was read in order to carry out the required actions. Because of working memory constraints, he or she will probably have to segment long sentences into smaller units before carrying out the actions (Boekelder and Steehouder 1999; Duggan and Payne 2001; Guthrie, Bennett and Weber 1991; Vermersch, 1985). After the actions have been carried out (interleaving reading and acting) it is necessary to make an assessment of the outcome. When the outcome seems correct (i.e., the performed procedure leads to the awaited result), the initial question can be considered as answered and the interaction with the text can stop. However, the outcome does not always fit the user's expectations. In this case, the user's interactions with the document continue. A possibility here is that the user restarts his/her interaction with the document from the beginning, rephrasing a question and searching for the corresponding information. Another possibility is that the reader will re-read the instructions to check that the formulated action plan was appropriate. If he or she thinks that everything was correct at this stage, he or she can reiterate the execution of actions until the awaited result is obtained.

4.2.4 Contributions and limits of this model

This three-stage model of user interaction with procedural documents shows that it is not only necessary to study the cognitive processes involved in reading the instructions, but also to study them before and after such reading. Indeed, the model can be used to orient design decisions intended to facilitate both the users' different types of activities and the cognitive processes implied in these activities (i.e., research, comprehension and application of the information). The document can be designed so that its structure can help users to locate the information. Its headings or its index can help them to compare this information with their goals, and its instructional format (text, picture, etc.) can help them to successfully integrate the necessary information.

5 An interactive model of the cognitive processes implied in the global and local processing of instructions

More recently, Ganier, Gombert and Fayol (2000) and Ganier (2004) have attempted to describe the behaviors and mental activities of adults learning to use a piece of equipment with the help of a procedural document. This document might be presented in texts, or pictures, or in a combination of the two. The model attempts to give an account of the different aspects of processing occurring at various levels: from the use of the document to the reading/understanding/execution of the instructions. In this model, that involves different informational resources (the equipment, the document's features and the user's prior knowledge), the interaction between the user, the document and the equipment does not start exclusively with reading of the instructions, but is heavily dependent on the setting of a goal representation either from the information displayed in the document or from the task to perform (see Figure 4).

The authors consider that the processing instructions involve eight different cognitive processes: (1) setting and holding a goal representation in work-

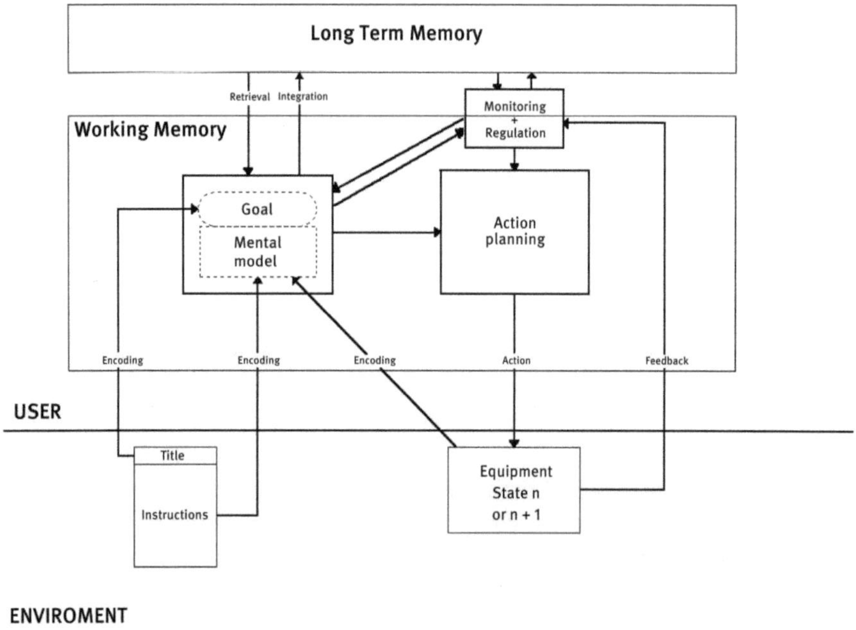

Figure 4: Ganier, Gombert and Fayol's (2000) model of processing of procedural instructions

ing memory; (2) encoding the instructions; (3) encoding the characteristics of the equipment; (4) integrating information from the document, the equipment and the user's prior knowledge; (5) action planning; and (6) executing the action plan. This model implies that (7) the user is able to monitor the procedure and compare the state of the equipment (the results of his/her actions) to the initial goal and (8) to regulate his/her activity until the achievement of the initial goal. According to this model, the processes occur in working memory which is constrained in both time and processing capacity (Baddeley 1986).

5.1 Setting and holding a goal representation

Setting a goal, holding it in working memory, and being able to recover it from long term memory are necessary processes and can be assumed as a first step when using a document to guide task performance. This first step can be done in different ways according to the situation, the user's prior knowledge, and his/her learning strategies. Thus, setting a goal representation will take place differently if the user is guided by the document rather than by the task. In the case of a beginner, or a cautious user carefully following the instructions prior to using the equipment, setting the goal representation might be done from reading the headings. In the case of an experienced user, or one guided by the task, setting the goal representation will correspond to his/her particular intentions: for example, "I would like to set this alarm clock for 7.30 a.m." Here, the user might start by using the document to look for information corresponding to his/her needs. In this case, the user will have to find a correspondence between his/her intentions and the representations induced by the headings or any other elements present in the document (for example, a picture or an index). In either case, the goal representation is held in working memory until the task is completed (Ganier 2004). This goal representation (either task- or instruction-based) allows the user to monitor his/her activity whilst performing the task. It is situated at a global level, constituting a kind of overall plan, allowing monitoring and regulation.

5.2 Encoding the instructions and integrating information from the document, the equipment and the user's prior knowledge

Using a procedural document to complete the goal set must lead to the construction of a representation of the to-be-performed actions (Mills et al. 1993).

According to Ganier, Gombert and Fayol (2000) and Ganier (2004), this second processing step involves integrating information from the instructions, the equipment and the user's knowledge. This implies processes guided by the document and the equipment on the one hand, and by the user's prior knowledge on the other. Integrating information from these different sources will lead to the elaboration of a mental model (that is a situational representation) that not only includes a representation of the actions to be performed but also of their localization and their modes of application (strength, direction, etc.). Manipulating this mental model allows the user to create an action plan for performing the action. Integrating these two different kinds of information presents two kinds of problem: (i) understanding the information contained in the instructions, and (ii) structuring the information in order to carry out the actions. Thus the speed, accuracy and effectiveness of elaborating and handling a mental model in order to carry out the actions is assumed to be dependent on the "quality" of the different informational resources (the instructions, the affordances of the equipment, and the user's prior knowledge) and, therefore, on the number of inferences that the user will have to make from these. Indeed, the risk is that the mental model proves to be incorrect or incomplete (and then ineffective) when one or the other of the informational resources is lacking, especially in the case of unskilled users (Ganier 2004).

5.3 Action planning and executing the action plan

The third step in the Ganier, Gombert and Fayol's (2000) model involves action planning. In order to perform the task, the user has to elaborate an action plan that specifies the order of the actions to be performed. This step consists of translating the declarative representation elaborated from the instructions, the equipment and the user's prior knowledge into a procedural representation aimed at performing the task. This step leads to carrying out the actions described in the instructions.

5.4 Goal-result adequacy monitoring and activity regulation

5.4.1 Monitoring the goal-result adequacy

Monitoring how far the results compare with the initial goal is necessary because of the need for accuracy. The model assumes that, whilst performing the task, the user compares the state of the equipment (i.e., following the results of his/her actions) to the initial goal. Thus, the effect of the user's actions are checked during – or immediately after – their execution. In order to do this, the user has to compare the goal representation in working memory to changes in the state of the equipment with reference to the instructions. Thus, the “reactions” of (or feedback from) the equipment constitute external indicators allowing the user to check if his/her interpretation of the instructions is acceptable and, in case of mismatches, to modify it (Richard 1990). A positive result leads to the next step. A negative result requires a return to a previous stage (with a re-reading of the instructions) either to manipulate the equipment by problem-solving or trial-and-error strategies, or to stop using it. Two consequences ensue from this monitoring process. First, in the absence of such a goal-result comparison process, mistakes leading to erroneous results are likely to occur. Second, such a monitoring process leads the user to regulate his/her own activity in the interaction with the document and the equipment.

5.4.2 Activity regulation

The goal-result adequacy monitoring process allows the user to control his/her activity until the initial goal is achieved. When the result matches the initial goal, processing of the current set of instructions can be followed by moving to a new set of instructions (corresponding to a new goal in the use of the equipment, or a new task that can be different from the current task). If, however, the result of the manipulations does not match the initial goal, the user can return to the instructions and process them again. If the instructions are still present in memory, a new execution phase can take place without the need to consult the document. If the instructions are judged insufficient, erroneous or incomprehensible, the user can try to reach the goal or solve these difficulties by problem solving or by trial-and-error strategies.

5.5 Integrating information in long-term memory

The acquired knowledge (consisting of a representation integrating the initial goal, the situation, the performed actions and the obtained result) is finally transferred via an integration process to long-term memory for permanent storage and later use. However, following Anderson (1983), Ganier, Gombert and Fayol (2000) and Ganier (2004) consider that the long term acquisition of procedures and the memorization of skills result from practice and experience and they, therefore, do not specify these aspects.

5.6 Contributions and limits of this model

One of the contributions of this model is that, on the one hand, it tries to identify what might constitute sources of difficulties for the users (see Ganier 2004). For example, it assumes that the efficiency – or the inefficiency – of the instructions can depend on one or other of the factors implied in the user-document-equipment interaction or on their combination. Thus, if one of these sources of information proves to be lacking (for example, when the instructions are inadequate, or the user lacks the required prior knowledge, or the equipment has no affordances to guide its use), then users have to try to solve these difficulties by themselves, and run the risk of failing to perform the task. On the other hand, identifying the users' cognitive processes should enable technical writers and document designers to develop document design solutions that fit the users' needs for information search and goal setting, mental model elaboration, action planning and activity monitoring (Ganier 2004; Ganier and Barcenilla 2007).

Turning to the limits of the Ganier, Gombert and Fayol (2000) model, the problem is, as with Bovair and Kieras (1991) and Guthrie, Bennett and Weber (1991), that this model focuses on how users read, understand and apply instructions from their first use of documents (i.e., the interpretative application of declarative knowledge) rather than on how such procedures are learned and stored in long term memory.

6 General discussion

The purpose of this chapter has been to present an up-to-date review of the models proposed in Cognitive Psychology to describe the activities and mental processes that are involved when people process procedural documents. These models can act as frameworks for the researcher, the teacher and the practitioner of technical writing or document design. For the researcher, they identify the variables at work in the processing of instructions, the understanding of text or pictures, and the learning of procedures. They also describe and lead to a better understanding of the roles played by different informational resources in this kind of situation. All of these elements can be utilized to formulate predictions about the effect of these factors on learning and comprehension. For the teacher or the trainer, these models can serve as educational tools. They can help technical writers to see the effects of (i) different ways of presenting information; and (ii) how some of the principles of technical writing can facilitate the use and understanding of such documents (e.g., why instructions should be presented in the form of numbered lists and not paragraphs, etc.). Because procedural texts are different from other kinds of texts, and because they implicitly specify ideal strategies for following them, the design and use of procedural documents could be specifically taught in classroom situations. They allow teachers to identify the sources of difficulties for users and teach how to avoid them (for example, teaching learners how to carry out step-by-step self-testing when carrying out procedures). For technical writers, these models can lead to a better understanding of the user and how to design such procedural documents or instructions based on knowledge gained from research rather than intuition (Ganier and Heurley 2005).

The development of such models is not only necessary from a theoretical point of view (allowing the development of knowledge about users' cognitive processing of procedural documents), but also from an applied and practical one. Using such models for technical training or at school would certainly not only improve the training of technical writers (aimed to improve document design), but also the training of (future) users. Both of these developments would lead to a better use and understanding of procedural documents.

Acknowledgement

The author would like to thank Professor James Hartley for his generous and friendly assistance in the preparation of this chapter.

References

Anderson, John R.
1983 *The architecture of cognition*. Cambridge, Massachussetts: Harvard University Press.

Alamargot, Denis, Chesnet, David, Dansac, Christophe and Ros, Christine
2006 Eye and pen: A new device to study reading during writing. *Behavior Research Methods, Instruments and Computers* 38 (2): 287–299.

Baddeley, Alan
1986 *Working memory*. Oxford: Oxford University Press.

Boekelder, Angelique and Steehouder, Michaël
1999 Switching from instructions to equipment: The effect of graphic design. In: Zwaga, Harm J. G., Boersema, Theo and Hoonhout, Henriëtte C. M. (eds.), *Visual Information for Everyday Use: Design and Research Perspectives*, 67–73. London: Taylor & Francis.

Bovair, Susan and Kieras, David
1991 Toward a model of acquiring procedures from text. In: Barr, Rebecca, Kamil, Michael L. , Mosenthal, Peter B. and Pearson, P. David (eds.), *Handbook of reading research, Vol. 2*, 206–229. White Plains, NY: Longman.

Dixon, Peter
1987 Actions and procedural directions. In: Tomlin, Russel S. (Ed.), *Coherence and grounding in discourse*, 69–89. Amsterdam/Philadelphia: John Benjamins Publishing Co.

Dixon, Peter, Harrison, Karen and Taylor, Dean
1993 Effects of sentence form on the construction of mental plans from procedural discourse. *Canadian Journal of Experimental Psychology* 47 (2): 375–400.

Duggan, Geoffrey B. and Payne, Stephen J.
2001 Interleaving reading and acting while following procedural instructions. *Journal of Experimental Psychology: Applied* 7 (4): 297–307.

Engle, Randall W., Carullo, Julie J. and Collins, Kathryn W.
1991 Individual differences in working memory for comprehension and following directions. *Journal of Educational Research* 84 (5): 253–262.

Ganier, Franck
2000/2001 Processing text and pictures in procedural instructions. *Information Design Journal* 10 (2): 146–153.

Ganier, Franck
2004 Factors affecting the processing of procedural instructions: Implications for document design. *IEEE Transactions on Professional Communication* 47 (1): 15–26.

Ganier, Franck and Barcenilla, Javier
2007 Considering users and their uses of procedural texts: A prerequisite for the design of appropriate documents. In: Alamargot, Denis, Terrier, Patrice and Cellier, Jean-Marie (eds.), *Improving the production and understanding of written documents in the workplace.*, 49–66. Amsterdam: Elsevier.

Ganier, Franck and Heurley, Laurent
2003 La compréhension de consignes écrites. In: Gaonac'h, Daniel and Fayol, Michel (eds.), *Aider les élèves à comprendre: du texte au multimédia,* 114–136. Paris: Hachette.

Ganier, Franck and Heurley, Laurent
2005 La prise en compte de l'utilisateur et de son utilisation des documents procéduraux: une précondition nécessaire à la conception de documents adaptés. In: Alamargot, Denis, Terrier, Patrice and Cellier, Jean-Marie (eds.), *Production, compréhension et usages des écrits techniques au travail,* 69–85. Toulouse: Octarès.

Ganier, Franck, Gombert, Jean-Emile and Fayol, Michel
2000 Effets du format de présentation des instructions sur l'apprentissage de procédures à l'aide de documents techniques. *Le Trav Hum* 63 (2): 121–152.

Glover, John A., Harvey, Anne L. and Corkill, Alice J.
1988 Remembering written instructions: Tab A goes into slot C, or does it? *British Journal of Educational Psychology* 58 (2): 191–200.

Guthrie, John T., Bennett, Stan and Weber, Shelley
1991 Processing procedural documents: A cognitive model for following written directions. *Educational Psychology Review* 3 (3): 249–265.

Johnson, Walter and Kieras, David E.
1983 Representation-saving effects of prior knowledge in memory for simple technical prose. *Memory and Cognition* 11 (5): 456–466.
Just, Marcel Adam and Carpenter, Patricia A.
1987 *The psychology of reading and language comprehension*. Newton, MA: Allyn & Bacon.
Kieras, David E. and Bovair, Susan
1986 The acquisition of procedures from text: A Production-system analysis of transfer of training. *Journal of Memory and Language* 25 (5): 507–524.
Kintsch, Walter and van Dijk, Teun A.
1978 Toward a model of text comprehension and production. *Psychological Review* 85 (5): 363–394.
Mills, Carol Bergfeld, Virginia A. Diehl, Deborah P. Birkmire and Mou, Lien-Chong
1993 Procedural text: Predictions of importance ratings and recall by models of reading comprehension. *Discourse Processes* 16 (3): 279–315.
Morrell, Roger W. and Park, Denise C.
1993 The effects of age, illustrations, and task variables on the performance of procedural assembly tasks. *Psychology and Aging* 8 (3): 389–399.
Richard, Jean-François
1990 Compréhension de textes à visée pragmatique. In: Richard, Jean-François, Bonnet, Claude and Ghiglione, Rodolphe (eds.), *Traité de Psychologie Cognitive 2: Le traitement de l'information symbolique*, 80–92. Paris: Dunod.
Schmid, Sabine and Baccino, Thierry
2001 Stratégies de lecture pour les textes à consignes. *Langages* 141: 105–124.
Sremec, B.
1972 Instructions, mechanical ability and perfomance. *Appl Ergon* 3 (2): 98–100.
Ummelen, Nicole
1997 *Procedural and declarative information in software manuals*. Amsterdam: Rodopi.
Van Dijk, Teun A. and Kintsch, Walter
1983 *Strategies of discourse comprehension*. New York: Academic Press.
Vermersch, Pierre
1985 Données d'observation sur l'utilisation d'une consigne écrite: L'atomisation de l'action. *Le Trav Hum* 48 (2): 161–172.
Wright, Patricia
1983 Manual dexterity: A user-oriented approach to creating computer documentation. In: *Human Factors in Computing Systems: CHI'83 Conference Proceedings*, 11–18. Boston: Ann Janda Editor.
Wright, Patricia
1999a Comprehension of printed instructions. In: Wagner, Daniel A. , Venezky, Richard L. and Street, Brian V. (eds.), *Literacy: An International Handbook*, 192–198. Boulder, Colorado: Westview Press.
Wright, Patricia
1999b Printed Instructions: Can research make difference? In: Zwaga, Harm J. G., Boersema, Theo and Hoonhout, Henriëtte C. M. (eds.), *Visual information for everyday use*, 45–66. London: Taylor and Francis.
Wright, Patricia and Wilcox, Penelope
1978 Following instructions: An exploratory trisection of imperatives. In: Levelt, Willem J. M. and Florais d'Arcais, Giovanni B. (eds.), *Studies in the perception of language*, 129–153. Chichester/New York: Wiley & Sons.

Sonja Ruda

How can the user be provided with the "right" questions and answers?

1 "Right" questions and answers

An instruction manual as a general description of how to use a product as well as a tutorial as a simulation of a specific task performance within a technical product provides the users with certain information. Initially, this information is imparted by an expert, here the developer of a product. This expert should be oriented towards the technical description, e.g., the product's construction, its components and their names, towards the service description, e.g., the product's purpose of use and the benefit for the user – and towards the operation description, e.g., the operations that must or can be carried out by the user. Moreover, the manufacturer of the product, a company, must be oriented towards certain issues regulated by legislation. This includes on the one hand warranty issues, i.e., the user's ability to set up or install and operate the product, and on the other hand issues of product liability, i.e., the prevention of potential hazards through warning notices (see Juhl 2005: 10 and 22–87). In other words, the expert is oriented towards these questions and supplies the corresponding answers. Technical editors, who write documentations about technical products, e.g. instruction manuals, thus having to deal with comprehensive communicative tasks (see Rothkegel 2010 and 2012/in this volume), support this process by suggesting questions to the expert and recording, sorting and formulating his answers (see Juhl 2005 and Grupp 2008).

Users are primarily interested in questions about the operation description and the operation instruction for particular operating elements of the product, which are tried to be resolved by consulting the technical manual and service descriptions.[1] The writer (experts and technical editors) of the instruction text is, amongst other things, confronted with the problem of determining the conceptual and local information relevance for the different users. For example if the help agent[2] within a text processing system asks whether the user would

1 The technical documentation of a product also includes technical details such as a circuit diagram, which are merely directed to technical staff. Since such details are irrelevant for the user of the technical product, they shall not be considered in this article. For the same reason no reference is made to the description of the functionality, which as a matter of fact is not a part of the instructions text (Juhl 2005: 91).

2 Help agents are generated dynamically as the system is keeping track of the user's behavior and actively corresponding to it (see Grünwied 2007: 43). Context-sensitive help is

like to write a letter although he is in the process of editing a scientific article or if in a telephone instruction manual the text passage on how to turn off the "annoying" speaker during a simple telephone call cannot be found. Some relevant issues are thus not presented at all or merely in an incomplete or ambiguous way. It is much too complicated to find them or they are identified wrongly by the online system. Practical applications of manuals reveal that "right" questions cannot be found by the writers on their own but only together with users (see, e.g., Nielsen and Loranger 2006).

The users do not necessarily have to be aware that they (should) ask more questions than those about the main operations with the appliance/software since they are simply not able to read the manual text if they proceed according to trial and error (see Ruda and Reißmann 2011 with regard to adolescents). The writers of an instruction text, however, have to consider and answer more questions than those about the main operations.

In the following, a model for determining "right" questions and answers as well as their definition and optimal arrangement shall be outlined. For that purpose it is necessary to simulate a conversation between experts, technical editors and users during the application of the product. Prior to this it is to be clarified in which form questions and answers are presented in instruction texts and what "right" questions and answers actually are.

1.1 Forms of questions and answers

Questions in instruction texts are first of all presented as interrogative sentences or keywords. They can be found in the table of contents, in headlines and in the index. Answers to the questions are given below the headlines and are presented as declarative sentences or, in case of instructions, as imperatives, e.g., "Select".

In order to simplify a "rapid" documentation, the propositions of questions and answers as well as their verbalizations are drafted from the writer's perspective (see tekom 2005) – even though the writers try to put themselves in the user's position and the user can be presented performing acts thus fading the writers into the background. However, as mentioned above, this writer's perspective can pose problems for the users. Hence "right" questions and answers have to be considered not merely from the writer's perspective but, above all, from the user's point of view.

---

offered, i.e., the help being dependent on the respective situation of the user (Grünwied 2007: 43 and 26).

1.2 What are “right” questions and answers?

Generally, “right” questions are on the one hand questions which are asked and answered by the writers of the instruction text when describing the setup and use of a product. On the other hand, “right” questions are primarily questions which are asked about the product by users – before, while or after reading an instruction text and using the product – and which yield “right” answers. In order to answer the question “What are right questions and answers?”, the following requirements can be considered:

Essential facts and acts: “Right” questions, which are asked by the writers and the user, are obvious, i.e., they necessarily have to be asked when using the product, e.g., where the power button of the TV set is located. “Right” questions focus on essential acts, e.g., how the TV set is switched on and how channels can be selected.

Coherent text structure: “Right” questions can be detected quickly through the indexes of the instruction text: they are formulated in a comprehensible way. Together with their answers they can be found in the main text and understood easily, i.e., in the interest of the user. “Right” questions and answers deal with the logic structure in the instruction text of the focused topic as well as completely listed subtopics and a comprehensible terminology.

Information relevance: “Right” questions are such that supply a satisfactory answer to the user through the instruction text. Satisfactory answers comprise propositions which provide the user with previously unknown information which is relevant to him. They are phrased in an easily comprehensible way. Whether an answer is satisfactory or not can on the one hand only be determined by the expert since only he possesses the technical background knowledge, which, however, raises the question to what extent the technical editor can act as an expert at all. On the other hand, it can only be the user who decides whether a text passage is incomprehensible, intricate or ambiguous.

Consideration of the expert’s and user’s perspective: “Right” questions and answers require an expert who first has to supply knowledge, which, through questions of users, can be completed or improved with regard to phrasing. The difficulty is to pool both perspectives, whereby the user perspective has to be in the center of attention for the text compilation, which is merely effected from the perspective of the expert or technical editor (see tekom 2005). Moreover, the writer has to face the difficulty of finding out where and how instruction texts can be improved.

Complying with laws: “Right” questions, which are asked by the writers, have to be specified by them or the manufacturer – e.g., questions about safety notes – since they are obliged to do so (see tekom 2005: 7–8). Question concepts referring to this do not necessarily have to be known to users, e.g., "Can the TV set be positioned anywhere in the living room?".

In conclusion, "right" questions are consequently essential questions about a product, i.e., compulsory questions, in other words, questions which in the first place permit the handling of the product and such which facilitate its handling or which aim at potential problems. The user must be able to find these questions easily in the form of headlines or in the table of contents or the index, i.e., questions must be adapted to the user's terminology. Likewise the writer's answers must be oriented towards the propositions of the questions and the user's terminology.

When determining relevant questions and corresponding answers, developers or their technical editors can describe the appliance or the software part by part: referring from the simplest to the most complex component, starting out from the global and afterwards going into details, explaining a process from the beginning to the end or including each module separately and having the user filter out the corresponding information through an index. This procedure involves the disadvantage that the potential user is not actually included thus possibly having other essential questions about the product. Ideally, an interested but still inexperienced user should try the product by means of a draft manual text. The question is which possibilities are suitable to "educe" questions from the expert and the user and how the experts' answers have to be formulated. It is imperative for both to cooperate, supported by a third person, a technical editor. There are different possibilities for this purpose, which shall be presented briefly in the following section.

2 Knowledge survey techniques for the determination of "right" questions and answers

As common for expert systems[3] (Puppe, Stoyan and Studer 2003: 607–608 and 610–611; Becker 1992: 50–56; Karbach and Linster 1990: 80–101; see also Ruda

3 Expert systems belong to the knowledge-based systems and thus to the main application areas of artificial intelligence. They are software systems storing the knowledge and problem-solving ability of experts and providing them to the user of an expert system (see Haun 2000: 143).

2008a: 27–32 and 2008b: 97–98), a number of knowledge survey techniques can be employed to determine "right" questions and answers.

2.1 Interviews

Interviews are suitable for gathering factual knowledge (declarative knowledge), i.e., knowledge about certain facts and circumstances and survey knowledge. A distinction can be made between the following types of interviews:

Structured/unstructured interview: A structured interview is an interview method that asks questions about a particular field of knowledge according to a particular schedule. Thematic branching is possible. It could be employed for questions about operations instructions. As a disadvantage unplanned fields of knowledge may be overlooked.

An unstructured interview uses an informal conversation to "educe" certain knowledge. However, it is problematic that important aspects may be forgotten and that, from a hypothetical point of view, implicit, unknown and vague knowledge cannot be investigated this way. The technical editor as an intermediary cannot scrutinize potential inaccuracies.

Focused interview: A focused interview aims at more profound knowledge about a topic. As in case of the above mentioned interview methods, it can be a disadvantage that the interviewer already needs to have substantial expert knowledge in order to be able to understand and appropriately process relevant facts.

Questionnaire: It is possible to employ questionnaires in interviews. The questionnaire, a research instrument of, e.g., psychology, contains direct questions to be answered by the test person as well as statements or keywords which the interviewee is to agree to or reject. The use of a questionnaire allows investigating correlations and dependencies of a limited number of characteristics (Mummendey 2003: 14–15 and 23). One problem is that the questionnaire cannot consider all relevant and previously unknown contents and possibly reveals vague knowledge thus showing a "measure-theoretical deficiency" (Mummendey 2003: 14). The exact theorization of practical acts – previously or subsequently – is to be considered critical as well. Another problem concerns the validity: Even if it can be assumed that questions are answered truthfully, real acts may deviate. Many acts depend on the specific context; they are unconscious, based on experience and also often too self-evident to be mentioned. This may constitute a disadvantage when realizing an operation instruction or solving a problem.

2.2 Expert observation

In an expert observation, the expert is recorded on video or audio tape while performing a problem solving process. The person recording the data must keep in the background. The expert observation mainly serves for gathering problem-solving knowledge. There are the following methods:

Introspection: Introspection means that the expert describes merely theoretically how he would solve a problem. Essential acts may be ignored for the instruction here.

Think aloud method: The think aloud method (van Someren, Barnard and Sandberg 1994) includes an expert solving a problem and, while doing so, verbalizing his thoughts. It may be criticized that the extent of the thought description is merely rough since the remarks lag behind the thoughts. It is impossible for the interviewee to verbalize and explain his complete considerations. Moreover, the verbalizations are concentrated on the analyzers (Puppe, Stoyan and Studer 2003: 610). Another problem involved in a process of thinking aloud is the lacking talkativeness, on the one hand referring to the quantity of remarks and, on the other hand, referring to the imprecise acts from the outsiders' point of view, e.g., while operating a product or during speech acts such as DESCRIBING[4] the product or EXPLAINING its functions. In such cases the person recording the data may ask again, however, if this occurs too frequently, the problem-solving process will be disturbed severely. In contrast, a too talkative expert can also cause difficulties. Becker (1992: 55) mentions the tendency to theorize as a problem even if there is no necessity to do so in the concrete application.

For the person recording the data it is often not easy to identify when an interposed question is indispensable. Since asking interposed questions too frequently may distract the expert, the recording person must completely keep in the background (Karbach and Linster 1990: 88). It is, however, difficult for the recording person to have open questions answered later. Therefore, rare interposed asking in a well-directed way does not necessarily have a disturbing effect (see Ruda 2008a: 32).

Unconscious acts are difficult to capture for both the person recording, since only the subsequent analysis can reveal where it would have been advisable to ask for implicit knowledge in more detail, and also for the experts if

4 For speech acts see Searle ([1969] 2007) and, for the context considered here, also Habscheid and Holly (2001). They are commonly quoted in capital letters.

they considered certain issues as too trivial to be worth explaining or if they did not know how to describe these issues, be it that they regarded the explanations to be sufficient or that they were simply not aware of it. Similar problems may occur among users if they do not or are not able to conceptualize the problem. This is particularly associated with the (mostly initial) fear of thinking aloud. This method is very suitable for having experts describe their own product, which is afterwards tested by users by means of a tutorial or an instruction text. The technical editor thus obtains detailed information on optimally or not (yet) optimally implemented passages.

Protocols: The think aloud method comprises two types of protocols:

(a) Concurrent protocols
Concurrent recording means that the expert solves a problem while thinking aloud. This results in a trace which, however, does not necessarily have to be complete and can be directly rounded off through a subsequent interview or, according to Ruda (2008a: 28), through the focused interview. This type of recording involves the advantage that all recorded acts were actually performed and that these acts, as well as unconsciously accomplished acts, can be exposed and revealed through a linguistic-pragmatic analysis (see Ruda 2008a: 143–206).

(b) Retrospective protocols
Acts described by the expert in retrospective protocols were not necessarily performed in reality since the problem solving is only specified subsequently. Information can thus be imprecise and essential questions may not be considered.

2.3 Indirect knowledge survey techniques

Some experts are not able to verbalize their knowledge through an expert observation. Furthermore, there are several knowledge areas that cannot be opened up that way. In order to attain indirect knowledge, the following techniques are employed (Karbach and Linster 1990: 92–101):

Multidimensional scaling: Multidimensional scaling is a knowledge collection technique where the expert evaluates the similarity of concepts by assigning them numerical values in a square matrix thus structuring these concepts. After these values have been recorded by the expert, analyzers form clusters of concepts. Multidimensional scaling seems appropriate if there are concepts which are structured but the principles of their organization are unknown or subjective.

Hierarchical clustering: Based on a similarity matrix developed by the expert, similarities between concepts (maximum, minimum, average) are gradually assigned in order to describe hierarchical structure. The two most resembling concepts from the original matrix are grouped. This group replaces both concepts and is newly included in the matrix. The entries for the new concept are subsequently updated. The pooling of two similar concepts and the regeneration of the matrix is continued up to the point where the matrix contains only one single group, i.e. the upmost hierarchy is reached.

Weighted nets: A similarity matrix is first developed into a coherent net (minimal connected network, MCN) in which the concepts are represented by the knots whereas the edges describe the weighting through the similarity evaluation between the concepts. Further connections are drawn in this graph if a shorter connection between two concepts can be found. This graph is called minimal elaborated network, MEN. Weighted nets intend to reflect the expert's associations between the concepts through a network.

Repertory grid: The technique of the repertory grid represents the expert's expectations towards occurrences in order to comprehend his differentiated decisions and acts. The repertory grid is based on the psychology of personal expectations according to Kelly ([1955] 1991). The psychology of personal expectations considers the human as a scientist who makes decisions on the basis of his expectations towards occurrences, which he comprehends through his constructs. In order to be able to understand a person's decisions, it is important to know the constructs that are used to describe his environment. According to Kelly's ([1955] 1991: 72) dichotomy corollary, the human construct system is made up of a limited number of dichotic (bipolar) constructs.

Structure laying techniques: When first analyzing an application field, numerous terms and concepts are elicited from sources like text books and interviews. Structure formation techniques are employed to determine first structures of the terms and the principles of the structure. In this process, terms and concepts are written on cards and subsequently ordered in groups. The expert is to arrange the cards of a group in such a way that the number of cards becomes as small as possible. The expert subsequently names the individual card groups. Next he is to combine the groups to larger groups and name them until no more groups can be formed. The expert's proceeding is recorded and analyzed by means of tape recording.

Closed curves: The technique of closed curves aims at identifying and marking related objects in physical rooms. Thus a chess player can mark logically linked pieces on the chessboard by drawing in closed lines around these chess pieces.

Inferential flow analysis: The inferential flow analysis intends to construct a causal network where the knots consist of concepts whereas the edges describe the kind and direction of the causality.

Ordered trees from recall: Here the expert enumerates ten to twenty times terms from a particular area whereby the first term always varies. Based on these enumerations it is tried to find terms (cores) which are always mentioned as coherent units in different enumerations. These cores are organized in a tree structure as leaves whose relations and hierarchies are determined subsequently. This technology distinguishes between the knowledge of experts and beginners and that of different experts.

These techniques of knowledge acquisition are also carried out automatically using tools (see Karbach and Linster 1990: 103–149) as in the case of ETS (Expertise Transfer System) (Boose 1986) – one of the first systems using the repertory grid procedure. The problem with indirect knowledge survey techniques is that they might be too abstract and thus too distant from the concrete case of application. Moreover, acts can be too complex to be ordered in a simplified manner through scaling methods. Indirect techniques of knowledge survey are not suitable for the consideration of an expert or users.

2.4 Content analysis and text analysis

Linguistically oriented methods of quantitative and qualitative content and text analysis (see Kuckartz 2009; Puppe, Stoyan and Studer 2003: 611–613 and Becker 1992: 56) are employed to “acquire subject-specific background knowledge” (Becker 1992: 51). Among them are the:

- keyword searching, where essential terms are filtered out,
- abstracting, i.e., the summary of basic contents of a (specialized) text and
- text mining, in which text contents are extracted, represented and made available again (see Mehler 2004a; Mehler and Wolff 2005; on quantitative methods Mehler 2004b).

It is also disadvantageous here that merely theoretical aspects can be filtered out. Therefore, these methods are not to be further considered for the presented topic.

In the following it shall be explored which knowledge survey techniques are suitable for experts and users in order to compose a good instruction text.

3 Knowledge survey: Expert interview – user interview

In the first instance it is essential to interview the expert, by means of which fundamental data and facts about the product are to be made explicit. Since an instruction manual is not regarded to be a specialist book for example engineers but rather a kind of textbook, practical guide or information brochure for users, the method of thinking aloud with the concurrent recording appears to be suitable. Here the expert can describe his product aloud by simply using it. The person recording the data, which should be done by the technical editor, can interfere in a supplementary way through the structured and/or focused interview. The collected data may be ordered thematically and described in more detail through indirect knowledge survey techniques. Furthermore, specialized texts could be analyzed with regard to the subject-specific background knowledge. After processing and analyzing the data, a textual and, where appropriate, a graphic or visual representation is to be prepared. Based on the data recorded on video, the expert and/or the technical editor compose the first version of an instruction manual, which is tested and evaluated with the product by the users. This, in turn, is effected through the method of thinking aloud.

Since the interest is laid on the detection and handling of potential problems through the users, the method of concurrent recording is to be preferred here as well. It may be supplemented by the structured and/or focused interview, which can solve potential ambiguities – while recording and/or in the process of screening the data. After a revision, the instruction manual should be tested by further users, which is again to be recorded on video, until all difficulties in using the product have been removed. The number of test users and test usages should be kept open in the first instance since it can only be fixed when the expert and/or the technical editor cannot detect new problems and previous and further test users can find their way through the instruction manual without any difficulties.

Priority is given to the detection and handling of problems through the users. According to Dörner (1987: 10), a problem is characterized by three components:

- undesired initial state s_α
- desired final state s_ω
- barrier momentarily preventing the transformation of s_α to s_ω.

The objective is to remove the barrier which prevents the transformation from the undesired initial state to the desired final state.

The problems found here are of the following kind: the expert solves unproblematic problems, the user unproblematic and problematic ones. Habscheid and Holly (2001: 185) ascertained these types of problems within the DFG project "The linguistic acquisition of computer media": If problems can be solved by means of "action routine" they are unproblematic problems, e.g., the expert detects and completes missing information while reading through the instruction manual, or the user is only able to perform an act successfully at the second attempt. Unsolvable cases, in contrast, involve real, i.e., problematic problems, e.g., if the user cannot switch off the speaker during a telephone call and is neither able to detect the corresponding button by simply trying nor by consulting the instruction manual.

Furthermore, there are problems on the part of the expert if he does not include propositions which are essential for the user or if he phrases them using expressions which the user is not able to find. These problems are to be conceptualized as real and equipped with the attribute unconscious, which makes them appear so difficult. The collected data is to be edited and analyzed, which shall be focused on in the following section.

4 Editing and analysis of data

The concurrent protocol of the expert and that of the first user are to be transcribed completely since all remarks may be important. A previous selection could ignore essential data. With regard to the additional user protocols it is sufficient to write down only new and problematic areas. It is recommendable to arrange the transcriptions in a table with each row having an empty row below for the notes of the analysis (see also Ruda 2008a: 318–323). Afterwards, the data is to be analyzed which results in analytical protocols. The later transcript analysis may again involve the problem of inaccuracy or even the loss of relations between facts. If necessary, the expert or the user is to be interviewed again. Suitable is an analysis that ascertains acts, e.g., the switching-on of an appliance, or speech acts (Searle [1969] 2007), e.g., CRITICIZING on the part of the user while using the product. It is fundamental in the first instance which information about the product has to be imparted by the expert, e.g., technical requirements, and which information can be imparted, e.g., special details. It is, above all, of primary importance that the user can cope with the expert's specifications. It is not only important "what" the user is doing but particularly "how" he is doing it. The transcript analysis involves the advantage that it can/should not only consider preset questions but also problems arising in the course of the analysis. The expert and/or the technical editor thus obtain a feedback, which can be used to critically improve the product description.

In order to ensure an optimal interpretation of data, product experts and technical editors should work closely with scientists of Applied Linguistics who have specialized in the field of language pragmatics and conversation research. Through the (language) pragmatic analysis of data, "right" questions and answers can be determined.

5 Specification of "right" questions and answers

Before finally being organized, the analyzed data is to be interpreted in terms of "right" questions and answers from both the expert's and the user's point of view. The classification can be taken from the transcripts or video recordings. A questionnaire can prove to be helpful here since merely a restricted number of classification possibilities can be inquired. Questions of the analysis can, for example, be:

(Absolutely) relevant information – less relevant information – users' state of knowledge

- Which information apart from legal requirements must, by all means, be part of the instruction manual?
- Which additional information can be included in the instruction manual?
- Which information is absolutely relevant? Which information is worth knowing?
- Which information is necessary for a basic usage, which for a detailed/ advanced one?
- Which information must be included in both versions?

Text layout

- Where is which information to be placed? The most important information should be at the beginning. Which order should be given to this important information?
- How can the user cope with the instruction manual's macro- and microstructure? Does it appear clear to the user? Is he able to find the quested points relatively quickly? Does he consider the instruction manual to be structured coherently? Can the most important information already be found at the beginning? Is it easily possible for the user to directly access particular details?

Text length
- Which text length is appropriate? Is the approach of "as much as necessary, as little as necessary" sufficient? Which concrete approach do the editors pursue when compiling the text?
- Is the text too short? Are there passages that are written too simple thus being too diffuse?

Text comprehensibility
- Are the texts comprehensible for the user?
- Are the terminology and the style appropriate for the user?
- Are technical terms explained? Is there a glossary for the user?
- Which technical terms are essential? Which ones can be omitted?

Structure and phrasing of the index
- Which phrases are to be chosen for the compilation of an index?
- Which synonyms are to be added?
- Are there sufficient references to other entries? Are all possibilities exhausted?

Use of illustrations
- Which illustrations are necessary?
- Do they assist the user in understanding the product better?
- Are they true to the original?
- Is a sketch more suitable?
- How are the illustrations to be described in the text? Is a short reference sufficient? Should the illustration be described in-depth?

Problem handling
- Which problems do occur? E.g., which functions cannot be managed by the user? Is the reason that the user cannot relate or implement the specifications in the instruction manual? Is the reason that he is not able to find the quested information in the text?
- How does the user try to solve these problems? Does he choose the trial and error method? Does he continue to search the instruction text for the wanted passage? Does he try to still understand the passage which he previously considered as incomprehensible? Does he consult other persons for help? Does he contact professionals?
- Are the user's problems solvable at all? Are his problems to be eliminated by simply correcting the instruction text? Or is it necessary to correct the appliance?

Which questions are relevant in the concrete case of application can only be determined through the user data.

6 Summary and outlook

This paper outlined a model for the determination of "right" questions and answers as well as their specification and optimal classification. The determination of these questions and answers should be effected through a conversation, a simulation, taking place between experts, the technical editor and users during the application of a technical product.

In order to specify "right" questions and answers for a product, they first of all have to be ascertained and eventually identified, analyzed and interpreted. An expert's knowledge, taken as the basis, is to be sequentially completed by the users' information up to the point where all considered users can cope with the instruction manual without problems. The person recording the data, preferably a technical editor, acts as intermediary. The knowledge survey can be based on the same techniques as for expert systems. Here, the method of thinking aloud, particularly that of concurrent protocols, is to be preferred since both the expert and the users disclose all thoughts about (speech) acts which are actually performed by them. Other methods such as questionnaires or retrospective protocols only allow a theorized subsequent description of (speech) acts and are thus never capable of presenting a complete trace. They may, however, provide useful additional information.

Each recording laid down in written form is analyzed with regard to acts, speech acts, problem indications and possibly immediate solutions, which significantly contribute to an improvement of the instruction manual. Starting out from the expert, questions about the product, the corresponding answers and their classification are compiled and, beginning with the second step, thoroughly completed with the users' acts, e.g., testing the product according to the instruction and evaluating the product manual in terms of text structure or formulations. Thus requirements for "right" questions and answers can be extended (see section 1.2) and questions can be concretized and expanded (see section 5).

According to Juhl (2005: 1) an ideal instruction manual cannot exist since users have a different existing knowledge and experience and instruction manuals cannot be considered as a substitute for a missing training. Nevertheless, the analysis of information from the user, above all using the method of thinking aloud, marks an essential step towards the objective of producing better user-oriented instructions.

Future studies on this model should include the determination of recurrent patterns of a product description, particularly with regard to problem determination and possible solutions, in order to facilitate the analysis of following similar products. Moreover, interview techniques can be simplified: questions of the focused interview can be gathered, ordered and thus provided to the expert or user in advance, which would result in fewer interruptions during concurrent recordings.

References

Becker, Barbara
1992 *Künstliche Intelligenz. Konzepte, Systeme, Verheißungen*. Frankfurt a.M./New York: Campus.

Boose, John H.
1986 *Expertise Transfer for Expert System Design*. (Advances in Human Factors/ *Ergonomics*) 3. Auflage. Amsterdam/Oxford/New York/Tokyo: Elsevier.

Dörner, Dietrich
1987 *Problemlösen als Informationsverarbeitung*. (Kohlhammer Standards Psychologie: Basisbücher und Studientexte) 3. Auflage. Stuttgart/Berlin/Köln/Mainz: Kohlhammer.

Grünwied, Gertrud
2007 *Software-Dokumentation. Grundlagen – Praxis – Lösungen*. (Kontakt & Studium 668.) 2., aktualisierte Auflage. Renningen: Expert.

Grupp, Josef
2008 *Handbuch Technische Dokumentation. Produktinformationen rechtskonform aufbereiten, wirtschaftlich erstellen, verständlich kommunizieren*. München: Hanser.

Habscheid, Stephan and Holly, Werner
2001 Sprechen vor dem Computer. In: Boehnke, Klaus and Döring, Nicola (eds.), *Neue Medien im Alltag: Die Vielfalt individueller Nutzungsweisen*, 176–195. (DFG-Forschergruppe "Neue Medien im Alltag") 2. Auflage, Lengerich/Berlin/Riga u. a.: Pabst Science Publishers.

Haun, Matthias
2000 *Wissensbasierte Systeme. Eine praxisorientierte Einführung*. (Edition expertsoft 36.) Renningen-Malmsheim: Expert.

Juhl, Dietrich
2005 *Technische Dokumentation. Praktische Anleitungen und Beispiele*. Illustriert von Tiki Küstenmacher. (VDI-Buch) 2., neu bearbeitete Auflage mit 135 Abbildungen. Berlin/Heidelberg/New York: Springer.

Karbach, Werner and Linster, Marc
1990 *Wissensakquisition für Expertensysteme. Techniken, Modelle und Softwarewerkzeuge*. München/Wien: Hanser.

Kelly, George A.
[1955] 1991 *The Psychology of Personal Constructs. Volume one: A Theory of Personality.* London, New York: Routledge 1991. Original edition, New York: Norton 1955.

Kuckartz, Udo
2009 *Einführung in die computergestützte Analyse qualitativer Daten.* 3. Auflage. Wiesbaden: VS Verlag für Sozialwissenschaften.

Lobin, Henning and Lemnitzer, Lothar (eds.)
2004 *Texttechnologie. Perspektiven und Anwendungen.* Tübingen: Stauffenburg.

Mehler, Alexander
2004a Textmining. In: *Lobin and Lemnitzer*, 329–352.

Mehler, Alexander
2004b Quantitative Methoden. In: *Lobin and Lemnitzer*, 83–107.

Mehler, Alexander and Wolff, Christian (eds.)
2005 Text Mining. *JLCL – GLDV-Journal for Computational Linguistics and Language Technology.* Volume 20 (1).

Mummendey, Hans Dieter
2003 *Die Fragebogen-Methode.* 4., unveränderte Auflage. Göttingen/Bern/Toronto/Seattle: Hogrefe.

Nielsen, Jakob and Loranger, Hoa
2006 *Web Usability.* München/Boston/San Francisco: Addison-Wesley.

Puppe, F., Stoyan, H. and Studer, R.
2003 Knowledge Engeneering. In: Görz, Günther, Rollinger, Claus-Rainer and Schneeberger, Josef (eds.), *Handbuch der Künstlichen Intelligenz*, 600–641. 4., korrigierte Auflage. München/Wien: Oldenbourg.

Rothkegel, Annely
2010 *Technikkommunikation. Produkte – Texte – Bilder.* (= UTB 3214.) Wien: Huter & Roth.

Rothkegel, Annely
2012 Communication on technology: Meeting point of theory and practice. In: *This volume.*

Ruda, Sonja
2008a *Aufgaben stellen, lösen und korrigieren. Eine sprachpragmatische Analyse für ein lehrerunterstützendes Feedback-Werkzeug im E-Learning.* Duisburg, Universitätsverlag Rhein-Ruhr. Zugl. Dissertation an der Technischen Universität Chemnitz, Philosophische Fakultät 2006.

Ruda, Sonja
2008b Model of a Teacher Assisting Feedback Tool for Marking Free Worded Exercise Solutions. *JLCL – Journal for Language Technology and Computational Linguistics* 23 (2): 96–108.

Ruda, Sonja and Reißmann, Kareen
2011 Sicherheitskommunikation über Transkriptanalyse und eine besondere Betrachtung der Nutzergruppe "Jugend". In: Banse, Gerhard and Krebs, Irene (eds.), *Kulturelle Diversität und Neue Medien. Entwicklungen – Interdependenzen – Resonanzen*, 315–322. (e-Culture/Cultural Diversity and New Media 16.) Berlin: trafo Wissenschaftsverlag.

Searle, John R.
[1969] 2007 *Sprechakte. Ein philosophischer Essay. Übersetzt von R. und R. Wiggershaus.* 10. Auflage Frankfurt a.M.: Suhrkamp. Original edition: Speech Acts. An Essay in the Philosophy of Language. Cambridge: University Press 1969.

van Someren, Maarten W., Barnard, Yvonne F. and Sandberg, Jacobijn A. C.
1994 *The Think Aloud Method. A Practical Guide to Modelling Cognitive Processes.* London: Academic Press.

tekom – Gesellschaft für technische Kommunikation e. V. (ed.)
2005 *Richtlinie zur Erstellung von Sicherheitshinweisen in Betriebsanleitungen.* Stuttgart.

Karl-Heinz Pogner

A social perspective on writing in the workplace: Communities of Discourse (CD) and Communities of Practice (CoP)[1]

1 Communities as contexts for text and knowledge production in the workplace?

"Community" has become a buzzword (and almost a magic spell) in literature on marketing and management. Numerous books and articles offer a wealth of advice on virtual communities to be assembled around a product or brand (cf. Andersen and Lindstrøm 1997: 195–216), on "rebuilding companies as communities" (Mintzberg 2009: 140), on commercial online communities to be developed (cf. Werry and Mowbray 2001: 1–125) and on professional communities of practice to be nurtured as a basis for creation, sharing and mediation of knowledge in organizations (cf. Wenger, McDermott and Snyder 2002, and others).

By this, the term "community", which has been familiar from academic literature with regard to text and knowledge production for quite a while now, has now also entered the discourse on business management. There might be a danger that the term "community" with its positive connotations could tempt people to confuse these communities with a kind of "cozy communitarian closeness" (Kreissl 2004: 37) and thus lose sight of conflict, criticism and power as components of these communities (Harris 1989). However, when we allow the term to include diversity and conflict, it can make a significant contribution to the analysis of organizations and the way they communicate and they create knowledge. In the sense of Taylor et al. (2001), for example, for which organizations consist of a network of "working communities" as the context for solving practical problems at the workplace and talking (and writing) about them. With regard to the computerization of labor, this means for example:

> Some communities (management) [...] take other communities (workers) as their object and relate to them by assuming they are a resource to be mobilized and structured, using as a tool the products of a third community ([IT] system design). Some communities ([IT]

1 This contribution has been pre-published as a working paper in English and in a German version (Pogner 2005 and 2007).

> system design) take the technologies of other communities (collaborative work) as their object and depend on the recruitment of a third community (management) to effect the implementation of their products into their work world (Taylor et al. 2001: 22).

Taylor et al.'s "working communities" can be characterized more precisely by two concepts taken from research into writing on the one hand and learning on the other, namely the concepts of discourse community and community of practice. In social-cognitive and social-interactive writing research, the concept of the "interpretation community", which derives from literary criticism, was taken up and developed further into the concept of the "discourse community" which has become popular in the research field of academic as well as non-academic writing. Research in the field of situated learning and social aspects of learning has developed the concept of "communities of practice" and very quickly applied it to the "learning organization". Furthermore, for a while now cultural and communication studies have been attempting to comprehend the phenomenon of "virtual communities" in the field of internet-based communication. However, it seems that no one has yet tried to examine these concepts of community in order to find out what they together can contribute to explore the social contexts and the social dimensions of text and knowledge production in organizations and to investigate the relationship between text and knowledge production. This paper is an attempt to fill this gap.

Therefore in the remainder of this paper, I will present the two central concepts of the discourse community (2.1) and the community of practice (2.2). Then, I will determine the specific contribution of these concepts for studying the social dimensions of text and knowledge production in organizations on the level of departments or project teams (2.3).

By way of three case studies, I shall then examine the explanatory value of the two concepts, which take the constituting and the constituted role of conventions, expectations and interaction in the communities mentioned into account (3). These case studies refer to internal and external communication at various Danish workplaces: text production by consulting engineers in the preparation of an energy concept (project team) (3.1), semi-public email communication at a university department (department) (3.2) and the work of two project teams developing IT solutions for a bank (project team) (3.3). The use of Information and Communication Technologies (ICTs) plays an important role in the last two case studies, therefore the concepts of virtual teams and distributed communities will be taken into account too. The empirical data for this examination of the two concepts' explanatory potential were collected in connection with case studies of my own carried out earlier (Pogner 2003, Pogner and Søderberg 2003) and with a master thesis (Business Administration) prepared under my guidance (Jensen 2002). The data consist of written documents (mainly in electronic form), qualitative interviews and ethnographic data such as semidetached or participating observations. The case studies draw on different approaches to analyze qualitative data: text analyses and analyses of text revisions (Pogner 2003), discourse analyses (Pogner and Søderberg 2003) or analyses of texts and interviews inspired by organization and learning theories (Jensen 2002).

The paper will conclude by discussing the contribution of the two concepts to the investigation of knowledge and text production (and of their possible interplay) in organizations (4).

2 Communities in text and knowledge production

Analyses of social networks have led us to realize that communities nowadays are not limited to immediate geographical neighborhoods or localities, and, inversely, that computer networks do not necessarily have to bridge long distances. Social networks facilitated by computer networks use different Information and Communication Technologies (ICTs: email, newsgroups, home pages, Internet Relay Chat (IRC), Web Logs (Blogs), Bulletin Board Systems, social networking platforms, Multi User Dungeons (MUDs and MOOs), video conferences, video streaming and other usually web-based technologies). The networks offer, at least potentially, the possibility of "companionship, social support, information and a sense of belonging" (Wellman and Gulia 1999: 169) – qualities which are characteristic for communities. Such virtual communities, in this case: communities linked by ICTs, create a common social reality in their interactive negotiating of identities, relationships, bonds and standards for the members (Bayum 1995).

Communities – regardless of whether the members are co-located or are connected "virtually" – are characterized by the development of certain behavioral rules and norms, but also by the feeling of belonging and engagement. That feeling is a result of shared ways of behaving and thinking – whether it is due to similar interests, tasks, professions or lots. These fundamental communal features and functions also characterize the concepts of discourse communities and communities of practice. In both cases knowledge is developed and shared. But the two concepts differ mainly in what forms the basis of the constitution of the community: the common discursive practice in solving knowledge problems or the everyday social practice via which mainly practical tasks are mastered. However, the one often includes the other – especially in work situations in which text production is an essential part of the tasks to be done (Jakobs 2005; Pogner 1999 and Couture 1992).

2.1 Discourse communities and text production

At the beginning of the eighties, the literary critic Stanley Fish demonstrated via the concept of interpretation community that the reception of texts is an active creative process on the part of the reader and that this process depends not only on the personal experience of the recipient but also on the conventions of the interpretation community to which he or she belongs (Fish [1980] 2000, 170–173).

> Interpretive communities are made up of those who share interpretive strategies not for reading (in the conventional sense) but for writing texts, for constituting their properties and assigning their intentions. In other words, these strategies exist prior to the act of reading and therefore determine the shape of what is read rather than, as is usually assumed, the other way round. (Fish [1980] 2000: 171).

"Writing" is used in the metaphorical sense here to describe the creation of the text by the reader in the reception process.

The concept of the interpretation community has been developed further to the concept of the discourse community in the context of social-cognitive and social interactive writing research. Discourse communities are characterized by the mutual dependence of language use and membership in the community:

> Discourse emphasizes that the group shares more than a particular native tongue or symbol manipulating skill. It connotes a complex set of conventions for assembling stretches of written or oral text, conventions shaped by cultural traditions as well as current circumstances. Community emphasizes that the people feel connected by virtue of their shared discourses and the work the discourse enables them to do (Bizzell 1994: 395).

The different discourse communities (family/home, school, workplace etc.) are characterized by various specific patterns of language use. They can be defined as social groups with common rules for language and its use, but also for the way of approaching problems. Members of the discourse community use their texts in order to demonstrate their membership in the community to which they belong or wish to belong (Pogner 2003; Winsor 1996).

Discourse communities affect the way in which their members define problems and formulate solutions, i.e., acquire, transform and produce not only language but also knowledge. The norms, conventions and expectations of a discourse community constrain the options of the members, but they also enable the communication of problem solutions and opinions. The influence goes in the other direction too: the members build up the community's framework of norms, conventions and expectations, reproduce and adjust it in their interaction or alter it if necessary. For this reason, besides the official and dominant forms of discourse there are also less official, alternative forms. Discourse communities are not necessarily harmonious, conflict free groups at all times, and they are not always free of the unequal distribution of discursive, social or economic power either. Academic discourse communities in particular live from discussion, criticism and academic competition in the name of knowledge creation – without being totally free of power constellations.

Academic research communities can be modeled as "specific interest groups" whose main purpose is to create or produce knowledge in a specific topic or subject area. They possess certain mechanisms and media which allow information to be exchanged and shared by their members, own a specific terminology and specific genres, which both include official texts (e.g., articles, conference papers, working papers) and unofficial, occluded texts (e.g., submission letters) (Swales 1990, 1996, 1998). In the official discourse the focus is often on argumentative texts which have to be based on well-documented proof. However, the positioning of one's own contribution within the discourse community is just as important as this rhetorical foundation.

The social constructivist or socio-interactive perspective and with it the potential field of application for the concept of discourse community is not limited to academic writing (cf. Bazermann 1988; Berkenkotter, Huckin and Ackerman 1991; Herrington 1985; Lehnen 2000). On the contrary, it has been expanded to include non-academic text production (cf. Killingsworth and Gilbertson 1992). In particular in the field of technical writing and especially research in professional communities such as engineering (cf. Pogner 1999, 2003; Winsor 1990, 1996; Zappen 1989) it has been used to explain individual and collaborative text production.

For writing in academic discourse communities, it is the case that text production not only accompanies social practices of problem-solving but itself often is an integral part of it (cf. the concepts of heuristic writing (Hermanns 1988) and of knowledge transforming (Bereiter and Scardamalia 1987)). Besides the shared language and discourse, academic discourse communities therefore also share practices of thinking, research and learning, which are expressed in the processes of changing, producing, disseminating and sharing knowledge in departments, institutes, universities, research associations, scientific disciplines and on conferences etc. Not all "social practices" (Fairclough 1992: 86–96) of research, information management, knowledge sharing, and learning are discursive in nature – but these "discursive practices" (Fairclough 1992: 73–86) are the starting point for the concept of the discourse community. Nowadays, many of these discursive practices take place in electronic newspapers, on electronic conferences, in electronic discussion lists, web logs etc. using ICTs.

Summing up, the characteristics of the discourse community can be described as follows:

	Aim	*Participants/ Members*	*Basis of cohesion*	*Duration*
Discourse community	Producing and disseminating knowledge in texts and discourses	No formal membership. Persons working on similar problems and with shared interests	Common (professional) interest. Participation in the joint discourse for problem-solving and for knowledge production.	As long as there is interest in the problem. Even if members often change, the DC continues to exist.

Table 1: Characteristics of discourse communities

2.2 Knowledge production and communities of practice

Social-psychological or social-cognitive researchers in organizational learning and the production, sharing and dissemination of knowledge are particularly interested in which social and organizational circumstances are beneficial for knowledge production. For researchers such as John Brown and Paul Duguid (1991) as well as Jean Lave and Etienne Wenger (1991), knowledge and learning are social phenomena and not rational, explicit "things" which can be produced in a formalized hierarchical work organization with comprehensive division of labor.

> Knowing is inherent in the growth and transformation of identities and is located in relations among practitioners, their practice, the artefacts of that practice, and the social organization and political economy of communities of practice (Lave and Wenger 1991: 122).

Organizations consist of complex relationships between different networks and groups which do not necessarily turn up in the organizations' charts but make sure that "the job is getting done".

> [...] A new organizational form is emerging that promises to complement existing structures and radically galvanize knowledge sharing, learning, and change. It's called the community of practice (Wenger and Snyder 2000: 139).

Communities of practice are groups of people who share an interest, a problem area or enthusiasm for a certain subject, look for problem solutions together and intensify their knowledge and expertise in this field via continuous interaction (Wenger, McDermott and Snyder 2002: 4).

> What are communities of practice? In brief, they're groups of people informally bound together by shared expertise and passion for a joint enterprise – engineers engaged in deepwater drilling, for example, consultants who specialize in strategic marketing, or frontline managers in charge of check processing at a large commercial bank. Some communities of practice meet regularly [...]. Others are connected primarily by email networks. [...] Inevitable, however, people in communities of practice share their experiences and knowledge in free-flowing, creative ways that foster new approaches to problems (Wenger and Snyder 2000: 139–140).

Knowledge creation and learning in these communities are characterized by sharing stories (the members of the community use these to diagnose problems and store the existing knowledge), cooperation (the members take part in a common practice) and a form of social construction of the community of practice (the members develop a common understanding of their practice and a general consensus as to how problems are to be solved) (cf. Brown and Duguid 1991).

Communities of practice come into being everywhere, and that includes different workplaces and companies but especially those organizations that want to/have to act in the context of the knowledge economy:

> We all belong to communities of practice. At home, at work, at school, in our hobbies – we belong to several communities of practice. And the communities of practice to which we belong change over the course of our lives. In fact, communities of practice are everywhere. [...] Workers organize their lives with their immediate colleagues and customers to get their job done. In doing so, they develop or preserve a sense of themselves they can live with, have some fun, and fulfil the requirements of their employers and clients. No matter what their official job description may be, they create a practice to do what needs to be done. Although workers may be contractually employed by a large institution, in day-to-day practice they work with – and, in a sense, for – a much smaller set of people and communities (Wenger 1998: 6).

Communities of practice combine learning in the sense of an activity (practice) with learning in the sense of belonging (community):

> [...] Practice: a way of talking about the shared historical and social resources, frameworks, and perspectives that can sustain mutual engagement in practice. [...] Community: a way of talking about the social configurations in which our enterprises are defined as worth pursuing and our participation is recognizable as competence (Wenger 1998: 5).

Common practice is the source of cohesion and membership and is characterized by the joint participation of the members in this practice. The practice is localized in the community of people and in their relations with each other which allow them to do what they do. However, what characterizes the community of practice on the inside is not only homogeneity (harmony, cooperation), but also variety (conflict, competition) – i.e., similar tensions as that are typical for discourse communities.

> [...] a community of practice can become a very tight node of interpersonal relationships. [...] These interrelations arise out of engagement in practice and not out of an idealized view of what a community should be like. In particular, connotations of peaceful coexistence, mutual support, or interpersonal allegiance are not assumed, though of course they may exist in specific cases. Peace, happiness, and harmony are therefore not necessary properties of a community of practice. Certainly there are plenty of disagreements, tensions, and conflicts [...] (Wenger 1998: 76–77).

Communities of practice are held together by negotiating the common goal or task (defines the domain of knowledge or practice), the mutual commitment (expresses how the community functions and what binds it together) and the common repertoire of the community (routines and skills, stories, styles, vocabulary, events and discourses (sic!)). Participation in communities of practice consists of common activity, interpersonal relationships, shared knowledge and negotiation of the shared interest (Wenger 2000: 230).

The properties of communities of practice can be summed up as follows:

	Aim	*Participants/ Members*	*Basis of cohesion*	*Duration*
Community of practice	Extension of the capabilities of the members. Exchange of knowledge	Members select themselves	Passion, commitment to and identification with the expertise of the group	As long as there is an interest in the continued existence of the group

Table 2: Characteristics of practice communities (cf. Wenger and Snyder 2000: 142)

2.3 Text and knowledge production in discourse communities and communities of practice

In order to define the terms used here more precisely, we shall now compare the concepts of the discourse community and the community of practice with traditional organizational units (for distinguishing between the community of practice and traditional units, cf. Wenger, McDermott and Snyder 2002: 43).

In contrast to departments (business units, functional or administrative units etc.), discourse communities and communities of practice are characterized by loose connections between the members and by a higher degree of informal self-management. Good relations among colleagues and participation are more important than official hierarchies. The main focus is on knowledge and learning, but there are still differences of power: old-timers have more power than novices; this power is based on their ability to contribute to the discourse and knowledge of the community. Discourse communities and communities of practice are able to transcend the boundaries of the organizations involved (cf. Brown and Duguid's notion of "networks of practice" (2000: 141–142)).

Characteristic for project teams is that they lead to the achievement of a common goal dictated from outside the team by carrying out interdependent tasks in division of labor. A project manager usually coordinates the individual efforts according to the goals of the team's task. In contrast, a coordinator of a community of practice is not really the manager of the community but the one who brings the members/participants together and allows them to find their own way. In discourse communities, there is usually no official manager or coordinator. The members of discourse communities or communities of practice are not interconnected via interdependent subtasks (like in a project team) but by interdependent knowledge and a common domain in which processes of knowledge sharing and learning take place. "[The community] is defined by its fundamental commitment to exploring its domain and to developing and sharing the relevant knowledge" (Wenger, McDermott and Snyder 2002: 43).

If we add the concept of discourse community to Wenger, McDermott and Snyder's (2002: 43) synopsis on community of practice, project team and department, the different structures can be summed up as shown in Table 3.

So what do communities of practice and discourse communities have in common and what distinguishes them from each other? They have in common the relatively voluntary nature of membership and the significance of the membership for the identity of the members. They also have in common the two constitutive elements "domain of knowledge", i.e., a set of topics and problems which the community is devoted to, and "community", i.e., the group of those interested in this domain (Wenger, McDermott and Snyder 2002: 27–40). But they differ in one other constitutive element ("practice" and/or "discourse"). If the focus is on the aspect of a common discourse developed in order to be effective in the domain in question, the community can primarily be analyzed as a discourse community. If the focus is on the aspect of common practice developed in order to be effective in the domain, the community can primarily

	Aim	*Participants/ Members*	*Basis of cohesion*	*Duration*
Discourse community	Producing and disseminating knowledge in discourses and texts	No formal membership. Persons working on similar problems and with shared interests. Fuzzy boundaries	Common (professional) interest. Participation in the joint discourse for problem-solving and knowledge production	As long as there is an interest in the problem. Even if members often change, the DC continues to exist.
Community of practice	Production, extension and exchange of knowledge, enlargement of the capabilities of the members	Members select themselves on the basis of expertise and interest in the subject Fuzzy boundaries	Passion, commitment and obligation; identification with the group and its expertise	As long as there is an interest in the continued existence of the group, in the subject and in learning
Project team	Execution of a specified task/ subtask	All who play a direct role accomplishing the task, assigned by management Clear boundaries	Milestones, overall goals and partial goals of the project	Until completion of the project
Department (functional or business unit)	Delivery of a product or duty	Everyone who reports (directly or indirectly) to the department's manager Clear boundaries	Job profile and requirements, common goals of department and organization	As long as the organizational structure in question continues to exist

Table 3: Comparison between different groups and communities in organizations

be examined as a community of practice. In some cases (for example in academic discourse communities), the community's practice consists mainly of official and unofficial discursive practices, i.e., discourse and communities of practice coincide here. In other cases, either practice is predominant (but this always includes language and discourse) or discourse (which is, however, always part of social practice).

If, for sake of argument, we replace "practice" by "discourse" in the following sentence, we get a good description of discourse communities:

> Whereas the domain denotes the topic the communication focuses on, the practice [or the discourse, khp.] is the specific knowledge the community develops, shares, and main-

> tains. When a community has been established for some time, members expect each other to have mastered the basic knowledge of the community [...]. This body of shared knowledge and resources enables the community to proceed efficiently with the domain (Wenger, McDermott and Snyder 2002: 29).

This substitution test demonstrates the close relationship between the two concepts. Both in discourse communities and in communities of practice domain-specific texts, discourses and a domain-specific language play an important role. However, with communities in which the discursive element is predominant, written texts occur more often than they do in communities in which the practical element and the spoken language are predominant. One of the instances in which the close relationship between the two concepts becomes visible, is the role which, according to Lave and Wenger (1991), language and discourse play in the socialization of newcomers to communities of practice:

> Indeed [...] learning to become a legitimate participant in a community involves learning how to talk (and be silent) in the manner of full participants. Talking within [a practice] itself includes both talking within [a practice] (e.g., exchanging information necessary to the progress of ongoing activities) and talking about [a practice] (e.g., stories, community lore). Inside the shared practice, both forms of talk fulfil specific functions: engaging, focusing, and shifting attention, bringing about coordination, etc., on the one hand; and supporting communal forms of memory and reflection, as well as signalling membership, on the other. [...] For newcomers then the purpose is not to learn *from* talk as a substitute for legitimate peripheral participation; it is to learn *to* talk as a key to legitimate peripheral participation (Lave and Wenger 1991: 105–109).

The same applies – to a probably even greater extent – for newcomers to discourse communities.

Discourse communities and communities of practice are thus not mutually exclusive concepts; they both describe the social aspects of knowledge creation (and learning) – with slightly differing priorities between discursive and non-discursive practices. The crucial question is: Does production of discourse, i.e., of coherent units of oral and written texts, serve as a means of knowledge production (as in a community of practice) or is text production also an important aim of knowledge production (as in a discourse community)? In the following case studies, it is the respective task of the producers of texts and knowledge, which is the main factor deciding which of the concepts mainly can be used to examine the social context of text and knowledge production. In the first case, the task is primarily the text production of a project team of engineering consultants, in the second it is the daily communication and working practice of a university department, and the third example deals with the development of IT systems in two project teams.

3 Case studies and examples

3.1 Text and knowledge production by a group of consulting engineers

The study of the genesis of an energy concept which Danish engineers prepared and composed for an East German town in the context of their consulting activities shows that the production of the various text versions, which build on one another, is embedded in chains of different interactions – some of them with German experts and clients on location (Pogner 1999, 2003). These chains of interactions, including text production, commentary and revision; but also visits, meetings, phone calls etc., allow the engineers to acquire local knowledge on the energy supply in the town in question and to use this knowledge as a starting point for finding technical, economical and environmentally sound solutions and for recommending and legitimating them in the energy-concept text.

On the one hand, the processes of text production and revision carried out by the Danish project team and the comments of the German clients and experts show that text production not only accompanies (non-linguistic) activity and problem-solving but itself is a form of interacting – and of solving problems. However, they also show that the text being produced and revised works as a catalyst for cooperation within the Danish project team and also with the German experts and clients. Furthermore the text (as an artifact) helps to focus meetings and discussions, contributes to organizing, planning and coordinating the task at hand, and it functions as a communication tool. These functions of texts and other artifacts are described by Kimble, Hildreth and Wright (2001: 231) as being typical for knowledge sharing in a community of practice when the members are not in the same place (co-located) but for example work in different countries (distributed). This indicates that communities of practice play an important role in the engineers' project.

The Danish engineers lack local information on the energy supply of German towns in general, the state of the power plants and supply networks in the town in question, and the potential cooperation partners available for the municipal services and their intentions and competencies. For this reason, they try to acquire the missing information by cooperating with local German experts – information which they could then transform into knowledge of their own via contextualization and evaluation (Diemers 2000: 371–372). On the one hand, they try to acquire "hard knowledge" (Hildreth and Kimble 2002) from other texts (maps, plans, drawings and diagrams, other expert reports, reports from visits, minutes from meetings etc.), and on the other hand they try to acquire "soft knowledge" (Hildreth and Kimble 2002) through their active cooperation with German experts and clients. The hard, declarative knowledge then can be articulated and codified, is learned via the processing, evaluation and revision of various documents. However, soft, non-codifiable, procedural knowledge, which is not so easy or impossible to quantify, grasp and store, can only be acquired in processes of participation (membership, activity, interaction etc.) and reification (shared documents, texts, instruments and other artifacts.), i.e., in two processes which are characteristic for communities of practice (Hildreth and Kimble 2002). At the end of the project, the Danish engineers concede self-critically that there was much too little concrete cooperation with the German experts and clients, i.e., that a community of practice with them was not established. The Danish engineers would like to have cooperated more closely with the experts and the clients. However, the way in which they carried out their task and organized the work on the energy concept ("via remote control from Denmark", as one of the engineers put it), made it impossible for a community of practice to evolve.

The challenges which the engineers face in interaction with the German experts and clients via different media, genres, and forms of communication (face-to-face, telephone, fax, email, discussion protocols, text drafts and versions, drawings, viewings, inquiries etc.) cannot only be explained by using the concept of the community of practice; some aspects can be dealt (better) with by resorting to the concept of discourse community. The job of the Danish project team is not just to design the energy supply of the German town with the help of the laws of thermodynamics and economic parameters while taking legal regulations and environmental considerations into account; it is also important for the consulting engineers to indicate how competent they are as experts through using the 'right' language and discourse. This is done by presenting oneself in texts and interactions as a member of the discourse community of engineers by displaying expert knowledge, using expert language and align with the "arhetorical" ideology of the production of technical texts (Winsor 1996) – even though one is aware that an energy concept does not only follow a technical and economic logic but political and strategic considerations, too.

Summing up, the challenges, difficulties and the partial failure of the Danish engineers can be explained with reference to the concepts of discourse community and community of practice. Only gradually did the engineers realize that the text to be produced was not only a text which had to indicate expertise by demonstrating membership in the discourse community of engineers but also a text which should have been directed to audiences characterized by other discourse communities (of economists, business people, investors, politicians, administrators and lawyers). To have to take all of this into account in a text written in a foreign language did not make the task any easier. Also, they did not succeed in building up a community of practice with those who not only speak the foreign language but also have the necessary local knowledge. The result was that only codifiable, explicit, hard local knowledge could be learned and used whereas it was not possible to acquire necessary, situated, procedural, soft local knowledge, meaning that only a limited amount of local knowledge was available for carrying out the planning and counseling task.

3.2 Computer-mediated communication at a University department

A second case study (Pogner and Søderberg 2003) focused on examining communicative practices at a Danish university department in the context of so-called "email-to-all"-communication and on the contribution of this form of computer-mediated communication to forming a community of the researchers/teachers and the administrative staff of the department. "Email-to-all"-communication allows every member of the department to send an email to all members of the department at once, thus allowing experience and information to be exchanged and shared and requests for assistance to be sent. In spite of this social potential, the "email-to-all" is not used to establish or support a community of practice. Either a community of practice of this type does not exist at all at the department, or it uses other media and channels. However, as a large portion of the researchers – in contrast to the administrative staff – do not so much consider the department as being a community tied to a certain physical location but rather as a kind of 'virtual community', whose members communicate often via telephone or email, the question arises as to why "email-to-all" is not used as a way of building up or supporting a community of this kind.

The senders in the email-to-all discourse are mainly managers and decision-makers; the topics deal with teaching, examinations, academic seminars and the organization of day-to-day work. The 'email-to-all' is used as a handy, quick 'mass medium' ensuring the practical broadcasting of information and decisions necessary to 'get the job done'. The discourse reflects the role distribution of the department: participation/membership is defined by being allocated to this department, the decisions are made and announced by the senders/management, and the recipients/employees are to carry them out in order that the department's products and services, i.e., research and, first and foremost, teaching, can be supplied.

One shows one's membership by paying attention to the conventions about what 'normally' is communicated (business matters), how this is communicated (factually, technically) and which genres and text types are used (announcements, messages, enquiries). A large part of the emails follows these discursive conventions; they can be interpreted as indicators for the existence of some sort of community. But no further signs of a sense of belonging to, identifying with, participating in the department as a community become visible in these 'typical' emails. If at all, a feeling of belonging becomes visible only indirectly in complains about the lack of community in those few mails, which are untypical for the email-to-all discourse of the department. The members of the department expressing themselves in these atypical mails are not managers or decision-makers. Other topics are dealt with via other genres and text types (narratives, stories, satire) in a different style (exaggeration, irony, parody): stories of "worst practices" in which everything goes wrong are told. Those stories are in sharp contrast to the "war stories" which Orr (1990, 1996) and Brown and Duguid (1991, 2000) see as an important medium of the expression of professional communities, in which pride, experience and expert knowledge and skills of a community are expressed. In the email-to-all stories, the department is constructed as the community of the employees, who are all in the same boat, which is characterized by stress, overtime and lack of support from managers and colleagues. Perhaps, this could be the starting point for a discussion on how this "in-the-same-boat" community could improve its practices in the direction of better practices so that less frustration occurs. However, a discussion of this kind does not take place in the email-to-all communication. We shall disregard at this point whether this discussion might be conducted in face-to-face communication between those who mainly see the department as being a unity fixed to a certain common location, i.e., a co-located community of practice. The observations mentioned confirm that "virtual teams" (Kimble, Alexis and Li 2000), "international communities of practice" (Hildreth, Kimble and Wright 1998 and Kimble, Hildreth and Wright 2001) or "electronic networks of practice" (Teigland and Wasko 2004) might neither be able to replace the entire face-to-face communication with computer-mediated communication nor can they disregard close interrelations between the physical world and the electronic world.

The department, which on the one hand shapes the email-to-all discourse and on the other hand is shaped or socially constructed, i.e., in this case: reproduced by it, is not enacted as an expert discourse community of researchers. This is not very surprising in view of the variety and at times fragmentation, which characterize this interdisciplinary department as a research unit. Neither does the department become visible as a community of practice, which could use the email-to-all as a forum for those who want to share knowledge and experience with regard to teaching and organizing teaching. The department is rather constituted as an administrative unit of the university which needs top-down communication in order to supply certain products and services: contributions to study programs in the form of teaching and examinations.

3.3 Knowledge production of project teams in a bank for the development of IT solutions

The last example deals with consequences of organizational changes caused by the merger of two Danish, a Swedish and a Norwegian bank, causing local communities of practice to be broken up (Jensen 2002) without being replaced for the purpose of the maintenance and improvement of (in particular soft) knowledge.

The restructuring of the organization means that the development of new IT products is reorganized. Development tasks used to be carried out by two cooperating project teams. One project team from the 'Business' sector which comprised the individual banks, had the role of an internal client which commissioned the other project team from the IT department to develop and implement a certain IT product. After the restructuring of the company, one single project team from the newly established 'Development and Organization' department carries out the IT development tasks (for example an IT system for the bank's pay desks or for credit cards). Besides IT employees, the new 'Development and Organization' department includes employees from the old 'Business' sector who used to be in charge of the individual banks and branches in Denmark, Sweden and Norway.

In consequence, information and knowledge sharing has not only to happen within the project team and between the "IT" and "Business" people represented in it. In order to prevent information and knowledge from being stored in 'information pockets' within a project team to which no-one outside of the project team has access, it is also necessary to ensure that information (as the input for knowledge creation and learning on the part of the 'learning organization') is made accessible to other project teams and the rest of the organization too. In everyday work, a large part of the sharing and disseminating of knowledge does not only take place in particular project teams but also in and between communities of practice in which knowledge is maintained and developed in collective activities. In contrast to project teams, they operate without formal management, deadlines or specific output (certain products or services).

The employees of the two project teams in the "Development and Organization" department notice that restructuring has made it harder to stay up-to-date with regard to knowledge of business processes. The members of the project team, who come from the old "Business" sector, miss the opportunity of direct sparring with the branches of the bank which allowed them to maintain their soft knowledge about business processes and to update it constantly. Furthermore, another organizational unit, "Credit Cards and Finances" used to be the context for a community of practice and thus for the maintenance of soft knowledge about business processes. Even after the merger, individual banks ask the former members of this unit for help. These continue to try to help the banks although they are officially no longer in charge and because the unit which is now responsible ("Bank Activities in Denmark") does not possess the necessary knowledge, because the employees who have built up this knowledge together were moved from the "Credit Cards and Finances" to the "Development and Organization" department or to the marketing department.

> The way it used to be, with all those people in the "Credit Cards" sector, we had everything within the card sector. [...] We used to have a box called "Cards and Finances" inside the company. But this was broken up and the people all ended up at different places [...]. We were all completely atomized. [...] This is a problem now, because we all used to be together, connected by the subject "card". Now that we have dispersed in all directions, we're fighting to preserve the card knowledge we had accumulated together. [...] I am no longer able to maintain and expand my knowledge. It is beginning to disappear because I don't have my fingers immediately in the matter any more (an employee of the project team quoted according to Jensen 2002: 60–61, original in Danish).

As a result, former employees of the old "Credit Cards and Finances" unit try to arrange informal meetings and stay in contact with each other in order to at least partially maintain the day-to-day sparring with others of like mind in the field – something which before the restructuring used to be quite normal. The management of the bank has taken no steps to encourage and support this initiative to maintain this community of practice. It has apparently not even noticed that there is a need for it after the restructuring. There is thus evidently a community of practice here which is lacking support – perhaps also electronic support.

With the exception of email, computer-mediated communication only makes a small contribution to information and knowledge sharing in the "Development and Finances" department. For the sharing of information and knowledge in communities of practice of the department, in particular socialization via 'apprenticeship', backup groups and a 'common language' as well as the abovementioned informal maintenance of knowledge in informal circles are seen as especially valuable and important. The contribution which computer-mediated communication makes to the work of the project teams and the maintenance of communities of practice is mainly to be found in the support of other means of internal and external knowledge sharing, means which are considered as more valuable. The concrete work of the project team produces common knowledge and also gradually functions as a “common interpretation space” (Diemers 2000: 369) which plays an important role in the contextualization, and thus the sharing of information and knowledge inside the team. Communities of practice can play an important role in making the soft, non-codifiable, procedural knowledge of the project teams accessible to other parts of the company. In the case study presented here, these communities emerge when old opportunities for exchanging experiences in co-located, official units disappear, for example as a result of restructuring.

The last case example demonstrates the effects of the breaking up of an organizational unit which formed the context for a community of practice: such a community of practice going diagonally to the project teams is sorely missed by the employees from the former ‘Business’ sector and is kept alive informally. This is not an easy task due to the lack of daily contact and togetherness and of official support by the management. The example also shows that, besides the relations prescribed in the official organizational charts, and in addition to the project teams set up by the management, there is a need for other forms of community, for example in the form of communities of practice.

4 Discourse and/or practice

In the case studies presented here, the aim was not to prove the ‘correctness’ of the concepts or to demonstrate the superiority of either of the two concepts. The purpose was to determine the degree to which the two concepts together can contribute to analyzing text and knowledge production in organizations and grasping in a more precise way the interactions between social relations, organizational context, communication processes as well as text and knowledge production. The case studies show that communities have a strong effect on how texts and knowledge are produced and passed on in organizations or departments and project teams within them – or at least how this could happen. The communities which can emerge within as well as between departments and organizations often have the characteristics of discourse communities and/or communities of practice, a fact which illustrates the social and interactive character of the production of texts as well as of knowledge in organizations.

In some cases it is difficult to make a sharp distinction between text production and knowledge production and between discourse community and community of practice – especially when the aim of the 'practice' of the community is the (in the widest possible sense) social or collaborative production of texts and discourses. The two types of community are important parts of the context of text and knowledge production and at the same time a result of them; but other contextual aspects can also shape membership and a sense of belonging (see for example the "all-in-the-same-boat" community of the university department). Furthermore, other aspects of the social and organizational context often have a similar strong effect as communities have (see for example the distribution of power and the division of labor in the university department).

In all three analyses, collaborative text production and/or other collaborative practices play a constituent role for the existence of communities which can promote knowledge production and learning in organizations.

- The first case (the energy concept of the engineers) shows that the project team must learn how to demonstrate membership of the discourse community of engineers and technicians while simultaneously showing consideration for other discourse communities (of politicians, business people, administrators) in order to be taken seriously as consulting engineers. It also shows how important communities of practice are in order to acquire soft knowledge necessary to solve planning and counseling tasks.
- In the second case (emails of a university department) the email-to-all discourse provides two different images of the department: the dominating image of a department which has a certain purpose to fulfill within the university, i.e., to provide teaching and research, and the much rarer image of a community of employees who are 'all in the same boat', and in which some employees indirectly complain about the lack of mutuality, common practice and a sense of belonging. The email-to-all communication is primarily aiming at spreading the information necessary for the functioning of the department and its 'production'. Voices expressing the lack of communal spirit have problems making themselves heard in this community. Email-to-all communication allows a lot of information to be passed on, but not much new knowledge is produced, and a sense of belonging becomes only visible in a negative way – if it becomes evident at all.
- The last example (project teams in a bank) also shows the need for communities of practice in order to be able to maintain and develop knowledge (soft knowledge in particular) without which the new 'Development

> and organization' department cannot execute its work in an optimum way. If they are not planned as part of the organization by the management, they arise nevertheless, thus confirming that membership in communities of practice is of a relatively voluntary nature.

All three examples show how important it is to be conscious of the potential of discourse communities and communities of practice for successful text and knowledge production. This also means recognizing them as important components of organizational contexts and promoting them if possible.

In the project teams and the department examined, ICTs (and email in particular) sometimes play an important role in addition to telephone and fax. In none of the examples, however, do purely virtual communities (in the sense of Rheingold 1993) occur. Teams and departments nevertheless temporarily function as such, i.e., without face-to-face interaction of the members. But in contrast to purely virtual teams or electronic networks of practice, the focus is on cooperation at a certain location (co-located communities); in phases in which this is not feasible, other means of communication are used as compensation. For processes such as planning and counseling, organizing work and production processes as well as coordinating text production or the production and sharing of knowledge, various different electronic media are used then – a characteristic strategy of geographically dispersed teams and communities.

At the end of this paper, I would like to raise the question, if it would make sense to transfer the concept of the community of practice to virtual communities (cf. Kimble, Alexis and Feng 2000). Whereas the concept of the discourse community easily can be expanded to take place in virtual and geographically dispersed communities, this is not so easy to do in the case of communities of practice. The attempt to transfer it to virtual teams has been criticized on the grounds that there is no shared practice in many cases – at least not a communal one and not one constituted in virtual space. In virtual space (e.g., in electronic newsgroups), the criticism goes, people report on a certain practice, discussing it and sharing common knowledge on it, but the practice itself is conducted outside of the virtual community. This at least is the way Christopher Lueg puts it in his analysis of the newsgroup "de.rec.bodyart", which deals with tattooing and piercing (Lueg 2000). This criticism can be countered in two different ways at the end of this paper. Either we expand the concept of practice so as to include discursive practice as a form of social practice, or we examine the newsgroup primarily as a discourse community whose common discourse is the end and the means of the community, i.e., the reason for its existence and cohesion. Whichever way is chosen, in the relevant newsgroup in which the virtual community comes together, the practice dealt with is not so much the practice of decorating the body but that of 'talking' (i.e., the discourse) about it.

Summing up, we can say that both discourse communities and communities of practice are important components of the social context of the production of texts and knowledge, even if they are not the only components. Expectations, practices and norms of the communities have an effect on the production of texts and knowledge in these communities – inversely, it is the common practice and/or the common discourse which makes the groups turn into communities.

In all three cases, two factors are decisive for the existence of the communities: firstly the participation of the members in a common practice, which is necessary in order that the feeling of belonging can arise in the first place, and secondly the consolidation of the practice in the form of norms, expectations and rituals – all signs of reification (Hildreth and Wright 2001: 231) of the practice into artifacts (for example narratives, written texts etc.). The three examples differ as to the way in which the common practice is primarily carried out. Does the community come into being during daily interaction and during the solving of practical problems, which for the concept of the community of practice are most important, or does the common practice consist of the production of texts and discourses, which does not only inscribe problem solutions and shared knowledge into texts, images, tables etc. but also itself produces problem solutions – and hereby knowledge?

The examination of the explanatory value of the concepts of discourse community and community of practice made in this paper shows that they can be used beneficially for the description of the discursive and social processes of text and knowledge production in organizations. It also confirms the close link between the two concepts which already was demonstrated in the theoretical section of the paper – especially with a broadminded understanding of "discourse" and "practice". For this reason, I believe one should take the influence and the constitution of both, the communities of practice and of discourse communities into account when analyzing text production, when analyzing knowledge production in organizations and when analyzing the interplay of the two production processes. Such analyses can then give information on whether texts are mainly there for reification purposes or whether text production is also the reason for the existence of the community. In the first case, the community would mainly be classified as a community of practice (with discourse as the means) and in the second case primarily as a discourse community (with discourse as the means and the end).

References

Andersen, Tim Frank and Lindstrøm, Martin
1997 *Mærkevarer på internettet.* København: Børsen.

Bazermann, Charles
1988 *Shaping written knowledge: The genre and activity of the experimental article in science.* Madison WI: University Press.

Bayum, Nancy K.
1995 The emergence of community in computer-mediated communication. In: Jones, Steven G. (ed.), *Cyber-Society: Computer-mediated communication and community,* 138–163. Thousand Oaks CA: Sage.

Bereiter, Carl and Scardamalia, Marlene
1987 *The psychology of written composition.* Hillsdale NJ/London: Lawrence Erlbaum.

Berkenkotter, Carol, Huckin, Thomas N. and Ackerman, John
1991 Social context contexts and socially constructed texts: The initiation of a graduate student into a writing research community. In: Bazerman, Charles and Paradis, James (eds.), *Textual dynamics in the professions: Historical and contemporary studies of writing in professional communities,* 191–198. Madison WI: University Press.

Bizzell, Patricia
1994 Discourse community. In: Purves, Allan C. (ed.), *Encyclopedia of English studies and language arts,* 395–397. New York: Scholastic.

Brown, John Seely and Duguid, Paul
1991 Organizational learning and communities-of-practice: Toward a unified view of working, learning, and innovation. *Organization Science* 2 (1), 40–57.

Brown, John Seely and Duguid, Paul
2000 *The social life of information.* Boston MS: Harvard Business School.

Couture, Barbara
1992 Categorizing professional discourse: Engineering, administrative, and technical/ professional writing. *Journal of Business and Technical Communication* 6 (1), 5–37.

Diemers, Daniel
2000 Information quality and its interpretative reconfiguration as a premise of knowledge management in virtual organizations. In: Malhotra, Yogesh (ed.), *Knowledge management and virtual organizations,* 365–379. Hershey PA & London: Idea Group.

Fairclough, Norman
1992 *Discourse and social change.* Cambridge: Polity.

Fish, Stanley
[1980] 2000 *Is there a text in the class? The authority of interpretive communities.* Cambridge MA/London: Harvard University Press.

Harris, John
1989 The idea of community in the study of writing. *College Composition and Communication* 40; 1, 11–22.

Hermanns, Fritz
1988 Schreiben als Denken. Überlegungen zur heuristischen Funktion des Schreibens. *Der Deutschunterricht* 4: 69–81.

Herrington, Anne J.
1985 Writing in academic settings: A study of the context for writing in two college chemical engineering courses. *Research in the Teaching of English* 19 (4), 331–359.

Hildreth, Paul and Kimble, Chris
2002 The duality of knowledge. *Information Research* 8 (1), paper 142 (http://InformationR.net/ir/8-1/paper142.html, 12. 04. 2005).

Hildreth, Paul, Kimble, Chris and Wright, Peter
1998 *Computer mediated communications and international communities of practice. Proceedings of Ethicomp '98*, 1998 March, Erasmus University, 275–286.

Jacobs, Eva-Maria
2005 Writing at work: Fragen, Methoden und Perspektiven einer Forschungsrichtung. In: Jacobs, Eva-Maria, Lehnen, Katrin and Schindler, Kirsten (eds.), *Schreiben am Arbeitsplatz*, 13–40. Wiesbaden: VS Verlag Sozialwissenschaften.

Jensen, Martin
2002 *Værdifuld videndeling – et casestudie i to projektgrupper.* (Unpublished Master thesis). Copenhagen: Copenhagen Business School.

Killingsworth, M. Jimmie and Michael K. Gilbertson
1992 *Signs, genres, and communities in technical communication.* Amityville NJ: Baywood.

Kimble, Chris, Hildreth, Paul and Wright, Peter
2001 Communities of Practice: Going virtual. In: Malhotra, Yogesh (ed.), *Knowledge management and business model innovation*, 216–230. London & Hershey PA: Idea Group.

Kimble, Chris, Alexis, Barlow and Li, Feng
2000 *Effective virtual teams through communities of practice. Management.* University of Strathclyde: Management Science Research Paper 2000/9.

Kreissl, Reinhard
2004 Community. In: Bröckling, Ulrich, Krasmann, Susanne and Lemke, Thomas (eds.), *Glossar der Gegenwart*, 37–41. Frankfurt a.M.: Suhrkamp.

Lave, Jean and Wenger, Etienne
1991 *Situated learning: Legitimate peripheral participation.* Cambridge: Cambridge University Press.

Lehnen, Katrin
2000 *Kooperative Textproduktion. Zur gemeinsamen Herstellung wissenschaftlicher Texte im Vergleich von ungeübten, fortgeschrittenen und sehr geübten SchreiberInnen.* (Ph.d. dissertation). Bielefeld: Universität Bielefeld, Fakultät für Literaturwissenschaft und Linguistik.

Lueg, Christopher
2000 Where is the action in virtual communities of practice? In: *Proceedings of the workshop "Communication and Cooperation in Knowledge Communities"* D-CSCW Conference: "Verteiltes Arbeiten – Arbeit der Zukunft", 11.–13. September in München.

Mintzberg, Henry
2009 Rebuilding companies as communities. *Harvard Business Review* 87 (7/8): 140–143.

Orr, Julian E.
1990 Sharing knowledge, celebrating identity: Community memory in a service culture: In: Middleton, David and Edwards, Derek (eds.), *Collective remembering*, 169–189. London: Sage.
Orr, Julian E.
1996 *Talking about machines: An ethnography of a modern job*. Ithaca NY& London: IRL Press.
Pogner, Karl-Heinz
1999 Schreiben im Beruf als Handeln im Fach (*Forum für Fachsprachenforschung 46*). Tübingen: Narr.
Pogner, Karl-Heinz
2003 Writing in the discourse community of engineering. *Journal of Pragmatics* 35, 855–867.
Pogner, Karl-Heinz
2005 *Discourse communities and communities of practice: On the social context of text and knowledge production in the workplace* (Working Paper 2005–080). Copenhagen: Copenhagen Business School.
Pogner, Karl-Heinz
2007 Text- und Wissensproduktion am Arbeitsplatz. *Zeitschrift Schreiben*, 11. 02. 2007 [open access: http://www.zeitschrift-schreiben.eu].
Pogner, Karl-Heinz and Søderberg, Anne-Marie
2003 Organisationsinterne 'E-Mail an alle'-Kommunikation: Informationsübertragung oder Kommunikationsraum? In: Handler, Peter (ed.), *E-Text: Strategien und Kompetenzen. Elektronische Kommunikation in Wissenschaft, Bildung und Beruf*, 235–254. (Textproduktion und Medien 7.) Frankfurt am Main: Lang.
Rheingold, Howard
1993 *The virtual community: Homesteading on the electronic frontier*. Reading MS: Addison Wesley, Reading.
Swales, John M.
1990 *Genre analysis: English in academic and research settings*. Cambridge: Cambridge University Press.
Swales, John M.
1996 Occluded genres in the academy: The case of the submission letter. In: Ventola, Eja and Mauranen, Anna (eds.): *Academic Writing: Intercultural and textual issues*, 45–48. Amsterdam/Philadelphia PA: Benjamins.
Swales, John M.
1998 *Other floors, other voices: A textography of a small university building*. Mahwah NJ: Erlbaum.
Taylor, James R., Groleau, Carole, Heaton, Lorna and van Every, Elizabeth
2001 *The computerization of work: A communication perspective*. Thousand Oaks CA: Sage.
Teigland, Robin and McLure, Molly Wasko
2004 Extending richness with reach: Participation and knowledge exchange in electronic networks of practice. In: Hildreth, Paul and Kimble, Chris (eds.), *Knowledge networks: Innovation through Communities of Practice*, 230–242. Harrisburg PA: Idea Group.
Wellmann, Barry and Gulia, Milena
1999 Virtual communities: Net surfers don't ride alone. In: Smith, Marc A. and Kollock, Peter (eds.), *Communities in Cyberspace*, 167–193. London: Routledge.

Wenger, Etienne
1998 *Communities of practice: Learning, meaning, and identity.* Cambridge: Cambridge University Press.
Wenger, Etienne
2000 Communities of practice and social learning systems. *Organization* 7 (2), 225–246.
Wenger, Etienne and Snyder, William M.
2000 Communities of practice: The organizational frontier. *Harvard Business Review* 78 (1), 139–145.
Wenger, Etienne, McDermott, Richard and Snyder, William M.
2002 *Cultivating communities of practice: A guide to managing knowledge.* Boston MA: Harvard Business School Press.
Werry, Chris and Mowbray, Mirinda (eds.)
2001 *Online communities: Commerce, community action, and the virtual university.* Upper Saddle River NJ: Hewlett-Packard, Prentice Hall PTR.
Winsor, Dorothy A.
1990 Engineering writing/writing engineering. *College Composition and Communication* 41 (1), 58–70.
Winsor, Dorothy A.
1996 *Writing like an engineer: A rhetorical education.* Mahwah NJ: Erlbaum.
Zappen, James P.
1989 The discourse community in scientific and technical communication: Institutional and social views. *Journal of Technical Writing and Communication* 19 (1), 1–11.

II. Information transfer through language(s) and pictures

Klaus Schubert

Technical communication and translation

1 Technical and specialized communication

Technical communication is the totality of a multitude of diverse, complex communicative acts. Many of them are carried out in a single language, while others make use of two or more languages. The latter comprise elements of translation, less frequently interpreting, or some other way of working in several languages and mediating between them. Technical communication studies and translation studies thus focus in part on the same objects, multilingual communicative acts of a technical content (cf. Schubert 2009).

This contribution is aimed at investigating the relationships between the monolingual and the multilingual elements of technical communication. In section 2, a four-dimensional model of technical communication is outlined which accounts for both the monolingual and the multilingual parts of the field. In section 3, I elaborate on the model showing the interdependencies and mutual influences between the four dimensions. Section 4 presents a feature analysis of the monolingual and multilingual elements of technical communication with a short conclusion in section 5. To set the scene for these analyses, I discuss a few definitorial features of the concept of technical communication in the following paragraphs.

Speakers of many languages find the English adjective technical to be ambiguous. On the one hand, the adjective can be used in its narrow sense to denote that which is related to engineering and technology and, on the other, in its broad sense to denote that which is related to any speciality. I take the adjective "technical" in the term "technical communication" in its narrow variant, with the broader field of communication on specialities of all kinds being called "specialized communication". Accordingly, technical communication is a subset of specialized communication. It is beyond the scope of this contribution to fully elaborate and theoretically underpin a comprehensive definition of either field. It may suffice here to mention the major features which in my opinion should be included in such a definition. These are:

- Technical communication is made up of oral or written communicative acts.
- The communicative acts are monolingual or multilingual.
- The communicative acts convey technical content.
- The communicative acts are of an informational kind.
- The communicative acts are part of the speaker's or writer's professional activity.
- Technical communication makes use of optimized means of communication.

These features delimit the realm of technical communication and set it off against neighbouring fields. The feature of technical content marks the difference from communication in specialities other than engineering and technology, for instance legal communication or medical communication. The feature of informational content tries to define the blurred division between technical communication on the one side and technical journalism and marketing communication on the other side. Technical journalism produces documents of an informational and opinionated nature, whereas marketing has a strong persuasive character, no matter how objective and informational the names of the marketing materials may sound ("product information"). The same feature also delimits technical communication from business communication. These divisions cut up a continuum of document types, so that there quite naturally are many documents which combine features of two or more of these fields, such as a contract (legal communication) which contains the technical specifications (technical communication) of an engine to be built and delivered or a customer information brochure (marketing communication) on a car which contains a technical data sheet (technical communication).

2 Four dimensions of technical communication

To account for the parameters which influence technical communication and other kinds of specialized communication, I have suggested a model of four dimensions (Schubert 2007: 248), which are

- the technical content
- the linguistic form
- the technical medium
- the work processes

of the communicative acts in specialized communication.

This set of dimensions is quite heterogeneous. Theoretical accounts of technical communication as well as analyses of communication at large have long been the virtually exclusive domain of linguistics. The fundamental concept of 20th-century linguistics is the linguistic sign combining content and form (Saussure [1916] 1969: 32). This underlying idea suggests a view on technical communication considering only the first two dimensions. In such a model, the technical medium of the communicative act is a mere accessory of the communication proper and plays no role worthy of scientific attention. In a similar way, the management of the work processes in which communicative acts are performed or documents are created and edited for use in communication seems not to be a worthwhile object of language- or communication-oriented study, though it of course figures in business administration studies. My four-dimensional model, however, is built on the diagnosis that the developments of the last fifteen or twenty years have changed the parameters of technical communication to such an extent that the dimensions of the technical medium and the work processes today exert an immediate influence on the linguistic form and the technical content of communicative acts in technical communication and vice versa. As a consequence, an analysis of technical communication today has to take all four dimensions into account. This implies broadening the perspective (cf. Schubert 2005) from a genuinely linguistic point of view to a wider, communicative approach.

Before applying the four-dimensional model to the monolingual and multilingual forms of technical communication, I shall clarify three concepts used in the preceding paragraphs, viz. the document, the technical medium and the work processes. In technical communication it is common to call the workpiece a document, rather than, as in linguistics, a text. I have tried to word what occurs to me to be a common understanding of the term as described in scholarly and professional publications as well as in standards and thus define a document as a fixed oral or written text, including nonverbal components, which can be reread or reheard at any time (Schubert 2007: 7). It is worth noting that along with written texts this definition comprises even oral ones, such as tape recordings, audio files, audio tracks of video tapes and DVDs, motion pictures, television programmes and the like, as long as they are fixed and replayable. Along with the text, the definition also includes nonverbal document components, such as graphics, photos, tables, diagrams and, in the case of electronic documents, animated graphics, audio and video sequences, computer programs, etc. The term *document* thus includes the text and everything which in a broad sense can be called illustrations. In addition, it also comprises the form and make-up of the text, thus in a written document the typography and the layout as well as the "functional" elements of the text, especially the access structure as established by outline, table of contents, index, bibliography, internal and external references, marginalia, etc.

Taking the document rather than the mere text as the primary workpiece in the communicative acts which are the object of study sets the scene for including the technical medium in the analysis. This term should therefore be established more clearly as well. All of the documents and some of the other communicative acts which do not produce anything with the status of a document have some technical shape or format. This includes print layout, computer file formats and others. I count these features as elements of the technical medium. The workpieces of technical communication are created and edited by means of tools, most frequently software systems. These are instruments of the work process and I thus subsume them under the dimension of the work processes. One could object that this distinction between the format of the documents (technical medium) and the software systems which output the formats (work processes) is artificial. Yet, I find it reasonable to differentiate between a process and its result because the same result may be the outcome of two quite different processes. For instance, a printed maintenance manual for a machine may be obtained by hard-formatting a word processor file or by merging text and format data in a content management system in the course of a single-source publishing process. The result is the same, but the processes are quite distinct and the possibilities for further use of the workpiece in onward work processes are much more varied in the latter case than in the former.

Having discussed basic concepts and definitions in a more comprehensive way, I now restrict the scope of analysis for the rest of this article to written technical communication only. Monolingual written technical communication is roughly identical to technical writing and the corresponding multilingual activity is technical translation. Technical writing is also called technical authoring. The professional community in some of the English-speaking countries, especially in the United States, increasingly prefers to call their speciality "technical communication". I choose to let "technical communication" remain the superordinate concept which includes technical writing, technical translation and their oral counterparts. Accordingly, I use "technical communicator" as a hyperonym for "technical writer" and "technical translator".

3 Interdimensional influences

The model of the four dimensions of specialized communication proposes to broaden our scholarly view on communicative work. This broadening is not merely desirable for a more complete picture of our object of study. My hypothesis is that new developments in the professional activities of technical writing and technical translation have made it necessary to include the two new dimensions in the analysis of technical communication. Obviously technical communication has always made use of some technical medium and the work processes have always been systematically organized in some way. Therefore, my hypothesis can only be sustained if it can be shown that the influence exercised by the two new dimensions is of a different quality rather than just of a higher quantity than before. With this objective, I shall in this section look into influences between the four dimensions, with particular attention to those exerted by the two new dimensions on the two established ones.

"Technical content" and "linguistic form". The influences from the dimension of the technical content on the dimension of the linguistic form are fairly traditional objects of study in terminology, lexicology, languages-for-special-purposes (LSP) research and translation studies. Among them is such a phenomenon as the vocabulary, especially the choice of terms, in a given technical document being determined by its subject field and the degree of speciality of a document determining the ratio between vocabulary from common and specialized language, respectively. Another case in point is the need of neologisms or new loan words caused by the novelty of the subject of a document. A third is the subject field-specific syntax which is a major object of study in LSP research.

"Linguistic form" and "technical content". An example of an influence exercised by the dimension of the linguistic form on the dimension of the technical content is found in cases where conceptual distinctions are triggered by linguistic distinctions. These phenomena are an instance of the effects of the language system on their speakers' world view studied by Wilhelm von Humboldt ([1820: IV 28] 2002: 19–20) and reworded in the Sapir-Whorf hypothesis of linguistic relativity (Sapir 1929; Whorf 1940).

These are some examples of influences between the two established dimensions of technical communication. They are the more traditional objects of research. The following paragraphs show the influences from the new on the established dimensions and between the new ones which motivate the four-dimensional approach. These are research objects which have come into focus more recently.

"Technical medium" and "linguistic form". As discussed above, the dimension of the technical medium comprises, roughly speaking, file formats and similar properties of documents. Influences from the dimension of the technical medium on the dimension of the linguistic form can be detected whenever the need to fit a given format determines a choice of words or style. A good example is found at present in the increasingly frequent situations in which the technical medium of documents dictates time and space limitations for the text to be written. These conditions occur for example in the subtitling of video sequences, television programmes, etc. Subtitling is a technique in which a video recording, motion picture, television programme or the like with moving pictures and a spoken dialogue or commentator text is furnished with an additional written equivalent of the spoken text. Intralingual subtitles are used for example for the hearing-impaired, interlingual subtitles are used for translation purposes. The wording of subtitles is heavily constrained by the available space, by the need to show the subtitles sufficiently long for reading, by the need to turn them off when their spoken equivalent is over and by the need not to let subtitles run across cuts in the film.

Of course video and television are mainly a matter of the mass media and the entertainment industry, but technically identical recordings are also used in technical communication, for example in multimedia manuals, and also in those types of document which lie in the overlapping fields between technical communication and the neighbouring areas of marketing communication and pedagogical communication, e.g., marketing-oriented product presentations, computer-based training materials, etc.

For the technical communicators, subtitling is especially tricky, since the dimension of the linguistic form is strongly influenced by both space and time constraints. In the core of the more traditional technical communication business, several fields of application are found where the linguistic form is "only" subject to space constraints. A typical example is the page-identical translation of documentation. This is a common requirement with user instructions, technical manuals, product fact sheets, etc. for which the customer provides source documents with a complex layout of a body of text interspersed with illustrations, tables, inset text boxes, etc. with the explicit condition that the target text should fit in exactly the same slots as the source text. The condition becomes still more difficult to fulfil, when the font and character size must also be identical with the source document. A similar case is documentation to be published in two or more languages in parallel columns.

A very large market segment in technical translation with an obtrusive influence from the dimension of the technical medium on the dimension of the linguistic form is software localization. Software localization is the translation of all text components of a software system with adaptations in functionality and wording made to meet the requirements of a new target market and with a validation process following the translation and revision work. Because of these additional work steps, software localization is often described by its representatives as an industry totally distinct from general technical translation, but if the nature of the work counts as evidence, software localization is technical translation work under special technical conditions and with a special element of adaptation. In software localization, all text components of a software product are translated. This includes the user interface, the online help, the online and the printed documentation, the texts on CD labels, packages, etc. The translation of online help documents has its special intricacies

(Heine 2006, 2010). In the present context, the translation of the text components of the user interface is the main issue. The translators' freedom of wording is heavily constrained by the space available on menu bars, pull-down menus, buttons, etc. The work process and the software tools used in localization normally provide for the possibility of enlarging the size of the functional elements of the user interface to accommodate target language forms which are longer than the source forms, but obviously there are narrow limits to such changes in the appearance of the user interface. The translators' efforts are still more constrained by some software developers' habit to present the text elements to be translated out of context, often in spreadsheet tables, and re-using single words or short phrases in different syntactic readings (e.g., "open" as an adjective, an infinitive and an imperative) while having them translated only once. The translation work is doomed to failure with almost total certainty, when system messages are built up by means of so-called concatenated strings. This is a way of hard-wiring the source language syntax into the program code of the system. This is a serious shortcoming in the original system design which cannot be remedied by a software localizer. The response to such counterproductive influences from the dimension of the technical medium on the dimension of the linguistic form is an influence in the opposite direction, to which I return below.

Other fields of technical translation with strong influences of the medium on the language, especially with space constraints, are the localization of the user interfaces of engines and similar technical products, the translation of the captions in technical drawings and the three-dimensional design graphics created by means of computer-aided design (CAD) systems, the translation of the documentation displayed in a data visor (Schneeberger 2003) or data helmet. Even the news services sending very short texts to personal digital assistants (PDA) or mobile telephones using the short-message system (SMS), at present primarily used in the entertainment and marketing industries, may within the foreseeable future be taken into use in technical communication – with all the space constraints they impose on the technical writing and translation.

The influences of the dimension of the technical medium on the dimension of the linguistic form discussed in the paragraphs above are caused by space and time constraints which condition linguistic choices by making the shortest wording the best one. There are at least two obvious corollaries from this type of influence: a counterinfluence of the language on the medium and an onward influence of the medium through the language on the content. I shall address these in turn.

"Linguistic form" and "technical medium". An influence of the dimension of the linguistic form on the dimension of the technical medium can be detected when the format of the documents is chosen or modified to fit the requirements of the language. Examples of such influences are translation-friendly document layout or software internationalization. A translation-friendly layout of documents to be submitted to page-identical translation, for instance, would leave empty space for allowing longer target texts. Internationalization of software design and development would mean to prepare the text components of software systems for translation, avoiding string concatenations, the re-use of words in different syntactic contexts or the embedding of text in graphics. (The issue of translation-friendly writing, which may come to mind in this context, is an influence of the linguistic form on itself.)

"Technical medium", "linguistic form" and "technical content". As mentioned above, the influence of the technical medium may not only condition the way something is worded but also what is said in the first place. If the influence goes this far, the technical medium influences the linguistic form and the technical content of the document. This is the case for example in subtitling, when the time and space constraints do not only push the translators to choose the shortest possible wording, but force them to simply omit parts of the content expressed in the spoken source text.

"Technical medium" and "work processes". Examples of influences of the technical medium on the work processes can be observed when a terminology database is used as a consistency tool. Terms provided in this form by a customer to a technical writer or technical translator will appear on the professional's screen. It is then recommended, or, depending on the exact stipulations of the contract, it may well be compulsory, to use these words. The technical communicators are in this way made to use words not chosen by themselves. This effect existed already in the pre-computer age in cases in which mandatory terminology was provided in lists typed on paper or, worse, hidden in reference documentation, but the quality of a term appearing as an answer to a query or by automatic term recognition in the communicator's target document is nevertheless of a different and considerably more compelling quality than before.

A very similar example of a qualitatively new obligation for the technical communicators is the use of translation memories. These systems contain previously translated sentences or sentence-like pieces of text in the source and the target language in parallel and become active as soon as the translator works on a sentence reasonably similar to one in the database. To what degree these memory sentences are mandatory for the translator again depends on the contract. It has, however, already become a habit in some market segments of technical translation to pay only for sentences not contained in the database, so that translators are not encouraged to do anything but accept the provided insertions in their work. Translation memory systems are a consistency tool much in the same way as term bases. They introduce into the work of the technical translators lexical, syntactic and stylistic choices not made by these professionals themselves. This kind of influence is by far more obtrusive than all that could be achieved by means of type-written term lists or the advice to stick to the language use of some reference documentation.

Above, the use of term bases and translation memories by a single translator was discussed with a view to the consistency of the current work with work done previously by the same or a different person. Term bases and translation memories, however, are also used as consistency tools in teams working simultaneously on different parts of a single large documentation. In such team settings the use of the software tools makes it necessary to organize the workflow in such a way that the tools can have their effect immediately with as little revision work as possible. This may for instance result in a team of translators working in a networked environment from a single server-based term base and into a single translation memory. This will allow for a wording chosen by one team member to become immediately available to all the others members, thus creating consistency by inserting wordings chosen by other team members. It goes without saying that such ways of propagating wordings are equally efficient in spreading good as well as bad translations. Networked teamwork may therefore call for additional provisions in the workflow to warrant quality – another indirect influence of the use of tools on the management of the work process.

Another qualitatively new influence of the technical medium is recombination. I use this term (Schubert 2003: 232) to denote the assembly of documents from pieces of text written in a different order and possibly in different contexts than the one in which they are read. Recombination texts are texts in which the reading path differs from the writing path (Schubert 2005). These texts are very common in technical communication, especially when hypertext technology is used such as in websites, online documentation, online help, etc. Recombination texts are also found in documents, with or without hypertext, which are composed from the contents of text block databases, for instance in document or content management systems. The effect on the text is normally a broken text-linguistic coherence. To some extent the same effect existed already in the pre-computer age in reference works, manuals and other documentation frequently used for nonlinear reading. The influence exerted by today's software tools could therefore be seen as quantitative rather than truly qualitative. However, it seems fair to assume (which, however, I have not verified by empirical data) that the opposite influence, from the linguistic form on the technical medium, to be discussed below, can counterbalance the influence under review here. This holds only in cases in which the professionals authoring the relevant documents have skills, time and authority to exercise this influence of the language on the medium. It is my assumption that these conditions much more often apply in the traditional production of reference works and similar types of documents than in today's hasty technical documentation preoccupied with cost reduction and resource re-use.

"Work processes" and "technical medium". An influence of the kind under review here is always likely to act in both directions. Accordingly, there is an influence of the dimension of the work processes on the dimension of the technical medium. Examples are work processes in technical communication that by their sheer complexity make it inevitable to use a workflow management system or workpieces which by their size and structure require the use of a document or content management system.

"Work processes", "technical medium", and "linguistic form". The influence of the work processes on the technical medium may have an onward effect on the linguistic form as well. When, for instance, a content management system with workflow functionality is used for the authoring and maintenance of larger volume product documentation, its various technical versions (e.g., print, online, web) and its parallel language versions, this is an immediate influence of the complexity of the work on the technical medium. Primarily, the tools of the work processes themselves are affected. Since a content management system allows for the smoothest workflows when other software tools needed by the technical communicators function as add-ins or utilities under the single interface of the content management system, the choice of software tools is conditioned by the choice of the content management system. The choice of tools implies those technical media to be chosen which can be created and edited by the selected tools. All tools together contain the consistency functionality discussed above in that they suggest to technical communicators pieces of text from the database of the content management system, and, possibly from the term bases and translation memories linked to it. This creates an onward effect on the linguistic form.

Viewing all these influences together we arrive at a totality of specialized communication composed of four closely interconnected dimensions. It is a challenging thought that these influences might not only be effective between, as it were, objective dimensions of workpieces being processed in anonymous workflows, but that they could be seen as effects exerted by the competence of the professionals carrying out the work processes under scrutiny here. I address this idea in a more comprehensive work (Schubert 2007).

4 Monolingual and multilingual technical communication

The analysis along the lines of the four-dimensional model does not differentiate between the monolingual and the bilingual activities in technical communication. In this section, I address the question whether and to what extent this undifferentiated view is justified, choosing as a starting point the overall work process. The work processes in technical communication form a sequence of three stages: production, translation and organization. In written technical communication, the typical workpiece is a technical document and the three stages are roughly identical to technical writing, technical translation and document management, respectively. The three stages depend on each other in the business administration sense of the term, since the output of one stage is the input for the next stage. A technical document will pass the first and optionally the second and/or the third stage in the indicated order. A document can be sent through some or all of the stages iteratively. I have previously suggested this three-stage view on technical communication (Schubert 2003: 226–228, 2005). While incorporating a broader range of objects, this view occurs to me to follow the line set out by Heidrun Gerzymisch-Arbogast and Klaus Mudersbach who postulate a description of the translation process as a sequence of steps which are dependent in the above sense (Gerzymisch-Arbogast and Mudersbach 1998: 30; Gerzymisch-Arbogast 2002: 23). The latter approach elaborates on the stance taken by a direction of translation studies which describes translation in a psychological way as a decision-making process (Jumpelt 1961: 169, 186; Levý 1967; Reiß [1981] 2000; Kußmaul [1986] 1994; Wilss 1988: 92–107, [1998] 2001; Shuttleworth and Cowie [1997] 1999: s.v. "Decision Making, Translation as"; cf. Gerzymisch-Arbogast 1997).

Technical writing and technical translation (plus document management, which I do not deal with in this article) are thus connected by the subsequent work on the same workpiece. This does not in itself imply any similarity of the nature of the work carried out in the two stages. However, most of the authors comparing the two activities (e.g., Lockwood, Leston and Lachal 1995: 63; Hartley and Paris 1996, 1997; Göpferich 1996, 1998: 10–11, 2002a, b, 2004: 23–25; Schmitt 1998; Sandrini 2002; Schubert 2003: 238–241, 2006b), find a great deal of common characteristics. The possibly most outspoken hypothesis is put forward by Peter A. Schmitt (1999: 25–34) who says that technical writing and technical translation are subsets of each other. The opposite pole in this spectrum is held by Gerhard Budin. Speaking not of the nature of the activities, but of the competencies of the professionals in technical writing and translation he finds them "quite distinct" (Budin 1994: 250) – Anthony Hartley and Cécile Paris (1996: 227, 1997: 114) disagree strongly.

In view of this range of opinions which (except for perhaps Budin) do not totally contradict each other but do not either give an unequivocal picture, it may seem appropriate to try to sharpen the analysis and establish more precisely which characteristics are shared by technical writing and technical translation and which distinguish the two fields. In my understanding, however, such an approach is not too revealing. Rather than looking into other and more finely differentiated characteristics, it appears to me worthwhile to define more carefully the logical status of the features taken into account.

Since the two fields are so similar in any objective analysis and still are subjectively felt to be distinct, I have suggested a feature analysis which gives up the intuitive assumption that two entities are best distinguished by features which one entity has and the other lacks. With the Prague School classification of logical oppositions in mind (Trubetzkoy [1939] 1989: 69), I choose four features which one of the fields has, while the other field may or may not. In other words, I investigate features which are constitutive for the one and occasional for the other activity (Schubert 2005, 2006a, b). The four features are

- interlinguality
- a source document
- researched information
- a change in audience design

I shall briefly comment on each feature.

"Interlinguality". This feature describes the rendering of a content from a source language into a target language. It lies in the dimension of the linguistic form. Translation is by definition interlingual. For technical translation, the feature is constitutive. Technical writing may or may not be interlingual. Normally, it is tacitly understood as a monolingual activity. However, as will be discussed below for the features of source document and researched information, technical writing is an authoring activity which makes use of a variety of sources of information. In professional practice it is very common that some of these sources are in languages other than the language of the document to be produced. The relationship between the information sources and the target document normally is not one of immediate translation but one of selecting and taking over portions of information rather than pieces of text. When foreign-language sources of this kind are used, the technical-writing work comprises interlingual steps. This is not, however, the case in all technical writing. Accordingly, technical writing as a whole may or may not be interlingual. For technical writing the feature is occasional.

"Source document". This feature describes the presence of a single document given at the onset of the technical communication activity with the stipulation that its content has to be rendered into the target document unchanged. This feature lies in the dimension of the technical content. Technical translation always starts from a source document. For technical translation, this feature is constitutive. Technical writing normally works from a variety of information sources, including development documentation, oral information from interviews with product developers and experts, and in some cases observational information from trial usage of the product. It is, however, possible that the content of some source documents is taken over into the target document literally or after some editing. In particular this is the case when subcontractor documentation is integrated into the documentation of a compound product. Therefore, for technical writing this feature is occasional.

"Researched information". This feature describes a working step in which the technical communicator retrieves information not contained in the source document (for technical translation) or the basic documentation (for technical writing) received from the customer. The feature lies in the dimension of the technical content. The additional information is used for understanding the source materials, for supplying a basis for making decisions of content or wording in the target document or for being directly included in the target document. In technical writing, this is an obligatory working step which draws on reference documentation received from the customer, published or other written materials from archives, libraries or the Internet, public or industry standards, developer and expert interviews and other sources. For technical writing, this feature is constitutive. In technical translation, researching additional information above and beyond the content of the source document is very often required, although there are source documents which are sufficiently self-contained. For technical translation this feature is therefore occasional.

"Audience design". The term *audience design* derives from Allan Bell (1984) and was introduced to translation studies by Basil Hatim and Ian Mason (1997: 12). I use it as meaning the adaptation of a text or document to the requirements of its readers in a broad sense which includes the content, the wording and the presentation of the document. The feature thus lies in the dimensions of the technical content and the linguistic form. It also extends into the dimension of the technical medium, since the appearance of a document contributes to the extent to which it is understandable or easy to use for a specific audience. My understanding of the term is that every document has some audience design which is mirrored in the author's deliberate or unconscious choices as to the linguistic and non-linguistic document components. Both technical writing and technical translation work with source materials, which all have their own audience designs. Technical writing works from a variety of sources into a single target document (or documentation) and it is an essential value added of this work to adapt the content, the language and the medium to the intended target audience. A change in audience design is therefore a constitutive feature of technical writing. For technical translation the requirements are different. The traditional condition is that the source document is to be rendered into the target language leaving everything but the language unchanged. This often implicit invariability requirement underlies many intuitive definitions of what translation is. It also is the basis of those directions of translation studies which speak of the equivalence of source and target in translation (introductory to the wealth of literature on the equivalence issue: Koller [1979] 2004: 159–300, 2004; Neubert 2004). However, this theoretical stance is frequently challenged resulting in a major debate, not least because quite frequently customers explicitly order a change in audience design as part of translation assignments. While these assignments may still be explained away by virtually dividing them into a translation job proper and a technical writing job, the theoretical debate has been fuelled even more by the observation that there are many communicative situations in which a change in audience design is a requirement inherent in the translator's task due to differences in knowledge, habits or expectations between the source and the target culture. What is discussed in translation studies under the headings of adaptation or explicitation is the finding that in these situations a syntactically or semantically equivalent translation does not exist because of intercultural differences, so that modifications in the wording and the content become inevitable. These modifications are changes in the audience design. They are by no means restricted to literary translation because cultural differences are relevant to technical translation as well (Baumann 1998). Since this is necessary in some, but not in all translation assignments, the feature of a change in audience design is occasional in technical translation.

The distribution of the four discussed features to technical writing and technical translation can be summarized as in the following table (Schubert 2006a). A plus marks a constitutive feature, a zero an occasional feature. None of the investigated features is permanently absent in any of the two fields. Such a feature would be marked by a minus.

Feature	*Technical writing*	*Technical translation*
Interlinguality	0	+
source document	0	+
researched information	+	0
audience design	+	0

5 Conclusion

The relationship between technical communication and technical translation, as analysed in this article, is at a superficial level a simple relation of inclusion, technical translation being a subset of technical communication. Any attempt, however, to define in terms of the nature of the work how technical translation is distinguished from the rest of technical communication, yields a complex picture. Technical translation is not, as it might intuitively seem to be, the bilingual or multilingual section of technical communication and technical writing is not the monolingual part. Rather, technical translation is that section in which the focus is predominantly on the interlingual mediation, while technical writing is that section in which the primary objective is adapting the audience design of a variety of sources to a single new target readership. The two sections of technical communication are not watertight compartments, but poles of a continuum. The professional practice has long discovered this, designing multifaceted working assignments and job profiles all the time. The emerging discipline of technical communication studies has yet to grasp the full variability of its object.

References

Baumann, Klaus-Dieter
1998 Kulturgebundene Determinanten der Fachübersetzung. *Zeitschrift für Angewandte Linguistik* 28: 34–52.

Bell, Allan
1984 Language Style as Audience Design. *Language in Society* 13: 145–204.

Budin, Gerhard
1994 New Challenges in Specialized Translation and Technical Communication. An Interdisciplinary Outlook. In: Snell-Hornby, Mary, Pöchhacker, Franz and Kaindl, Klaus (eds.), *Translation Studies. An Interdiscipline*, 247–254. (Benjamins Translation Library 2.) Amsterdam/Philadelphia: Benjamins.

Gerzymisch-Arbogast, Heidrun
1997 Translating Cultural Specifics: Macro- and Microstructural Decisions. In: Hauenschild, Christa and Heizmann, Susanne (eds.), *Machine Translation and Translation Theory*, 51–67. (Text, Translation, Computational Processing 1.) Berlin/New York: Mouton de Gruyter.

Gerzymisch-Arbogast, Heidrun
2002 Ansätze der neueren Übersetzungsforschung. In: Best, Joanna and Kalina, Sylvia (eds.): *Übersetzen und Dolmetschen. Eine Orientierungshilfe*, 17–29. (UTB 2329.) Tübingen/Basel: Francke.

Gerzymisch-Arbogast, Heidrun and Mudersbach, Klaus
1998 *Methoden des wissenschaftlichen Übersetzens.* (UTB 1990.) Tübingen/Basel: Francke.

Göpferich, Susanne
1996 (Fach)Übersetzen vs. Technical Writing: Parallelen und Unterschiede. In: Budin, Gerhard (ed.), *Multilingualism in Specialist Communication/Multilingualisme dans la communication spécialisée/Mehrsprachigkeit in der Fachkommunikation, Vol. 1*, 399–434. Vienna: TermNet.
Göpferich, Susanne
1998 *Interkulturelles Technical Writing*. (Forum für Fachsprachen-Forschung 40.) Tübingen: Narr.
Göpferich, Susanne
2002a Neue Wege zu multilingualer Dokumentation: Globalisierung und ihre Rückwirkung auf die Berufsfelder Technische Redaktion und Übersetzen. In: Zybatow, Lew N. (ed.), *Translation zwischen Theorie und Praxis*, 167–189. (Forum Translationswissenschaft 1.) Frankfurt am Main etc.: Lang.
Göpferich, Susanne
2002b Vom Übersetzen und der Technischen Redaktion zur Interkulturellen Technischen Redaktion. In: Feyrer, Cornelia and Holzer, Peter (eds.), *Translation: Didaktik im Kontext*, 105–121. (InnTrans. Innsbrucker Beiträge zu Sprache, Kultur und Translation 1.) Frankfurt a.M. etc.: Lang.
Göpferich, Susanne
2004 Die Fähigkeit zur übersetzungsgerechten Textproduktion als Teil der Translationskompetenz. In: Fleischmann, Eberhard, Schmitt, Peter A. and Wotjak, Gerd (eds.), *Translationskompetenz*, 353–376. (Studien zur Translation 14.) Tübingen: Stauffenburg.
Hartley, Anthony and Paris, Cécile
1996 Automatic Text Generation for Software Development and Use. In: Somers, Harold (ed.), *Terminology, LSP and Translation*, 221–242. (Benjamins Translation Library 18.) Amsterdam/Philadelphia: Benjamins.
Hartley, Anthony and Paris, Cécile
1997 Multilingual Document Production: From Support for Translating to Support for Authoring. *Machine Translation* 12: 109–128.
Hatim, Basil and Mason, Ian
1997 *The Translator as Communicator*. London/New York: Routledge.
Heine, Carmen
2006 Herausforderung Hypertextübersetzung. In: Heine, Carmen, Schubert, Klaus and Gerzymisch-Arbogast, Heidrun (eds.), *Text and Translation: Theory and Methodology of Translation*, 17–39. (Jahrbuch Übersetzen und Dolmetschen 6.) Tübingen: Narr.
Heine, Carmen
2010 *Modell zur Produktion von Online-Hilfen*. (Forum für Fachsprachen-Forschung 90.) Berlin: Frank & Timme.
Humboldt, Wilhelm von
[1820] 2002 Ueber das vergleichende Sprachstudium in Beziehung auf die verschiedenen Epochen der Sprachentwicklung. – new edition: Humboldt, Wilhelm von (1963): *Werke in fünf Bänden*. In: Flitner, Andreas and Giel, Klaus (eds.), Vol. 3: *Schriften zur Sprachphilosophie.*. 9th ed. 2002. Darmstadt: Wissenschaftliche Buchgesellschaft.
Jumpelt, Rudolf Walter
1961 *Die Übersetzung naturwissenschaftlicher und technischer Literatur*. (Langenscheidt Bibliothek für Wissenschaft und Praxis 1.) Berlin: Langenscheidt.

Koller, Werner
[1979] 2004 *Einführung in die Übersetzungswissenschaft*. 7th revised ed. 2004. Wiebelsheim: Quelle & Meyer.
Koller, Werner
2004 Der Begriff der Äquivalenz in der Übersetzungswissenschaft. In: Kittel, Harald, Paul Frank, Armin, Greiner, Norbert, Hermans, Theo, Koller, Werner, Lambert, José with House, Juliane and Schultze, Brigitte (eds.), *Übersetzung/Translation/Traduction*, Vol. 1, 343–354. (Handbücher zur Sprach- und Kommunikationswissenschaft 26.1.) Berlin/New York: de Gruyter.
Kußmaul, Paul
[1986] 1994 Übersetzen als Entscheidungsprozeß. Die Rolle der Fehleranalyse in der Übersetzungsdidaktik. In: Snell-Hornby, Mary (ed.), *Übersetzungswissenschaft – eine Neuorientierung*, 206–229. 2nd ed. 1994. (UTB 1415.) Tübingen/Basel: Francke.
Levý, Jiří
1967 *Translation as a Decision Process. To Honor Roman Jakobson*. Vol. 2. (Janua linguarum, Series maior 32.) The Hague etc.: Mouton, 1171–1182.
Lockwood, Rose, Leston, Jean and Lachal, Laurent
1995 *Globalisation: Creating New Markets with Translation Technology*. London: Ovum.
Neubert, Albrecht
2004 Equivalence in Translation. In: Kittel, Harald, Paul Frank, Armin, Greiner, Norbert, Hermans, Theo, Koller, Werner, Lambert, José with House, Juliane and Schultze, Brigitte (eds.), *Übersetzung/Translation/Traduction*, 329–342. Vol. 1. (Handbücher zur Sprach- und Kommunikationswissenschaft 26.1.) Berlin/New York: de Gruyter.
Reiß, Katharina
[1981] 2000 Type, Kind and Individuality of Text: Decision Making in Translation. *Poetics Today* 2 [4]: 121–131 – Reprinted in: Lawrence Venuti (ed.), (2000): *The Translation Studies Reader*, 160–171. London/New York: Routledge.
Sandrini, Peter
2002 Mehrsprachige Fachkommunikation: Wissens- und Kulturtransfer im Zeitalter der Globalisierung. In: Zybatow, Lew N. (ed.), *Translation zwischen Theorie und Praxis*, 395–410. (Forum Translationswissenschaft 1.) Frankfurt am Main etc.: Lang.
Sapir, Edward
1929 The Status of Linguistics as a Science. *Language* 5: 207–214.
Saussure, Ferdinand de
[1916] 1969 *Cours de linguistique générale*. In: Bally, Charles and Sechehaye, Albert with Riedlinger, Albert (eds.). New ed. 1969. Paris: Payot.
Schmitt, Peter A.
1998 Technical Writing und Übersetzen. In: Snell Hornby, Mary, Hönig, Hans G. , Kussmaul, Paul and Schmitt, Peter A. (eds.), *Handbuch Translation*, 154–159. Tübingen: Stauffenburg.
Schmitt, Peter A.
1999 *Translation und Technik*. (Studien zur Translation 6.) Tübingen: Stauffenburg.
Schneeberger, Josef
2003 Die Doku in der Brille. Ein Blick in die virtuelle Welt. *technische kommunikation* 25 [3]: 34–39.

Schubert, Klaus
2003 Integrative Fachkommunikation. In: Schubert, Klaus (ed.), *Übersetzen und Dolmetschen: Modelle, Methoden, Technologie*, 225–256. (Jahrbuch Übersetzen und Dolmetschen 4/I.) Tübingen: Narr.

Schubert, Klaus
2005 Translation Studies: Broaden or Deepen the Perspective? In: Dam, Helle V., Engberg, Jan and Gerzymisch-Arbogast, Heidrun (eds.), *Knowledge Systems and Translation*, 125–145. (Text, Translation, Computational Processing 7.) Berlin/New York: Mouton de Gruyter.

Schubert, Klaus
2006a Interkulturalität in technischer Redaktion und Fachübersetzung. In: Wolff, Dieter (ed.), *Mehrsprachige Individuen – vielsprachige Gesellschaften*, 191–204. (Forum angewandte Linguistik 47.) Frankfurt am Main etc.: Lang.

Schubert, Klaus
2006b Kultur, Translation, Fachkommunikation. In: Heine, Carmen, Schubert, Klaus, Gerzymisch-Arbogast, Heidrun (eds.), *Text and Translation: Theory and Methodology of Translation*, 391–409. (Jahrbuch Übersetzen und Dolmetschen 6.) Tübingen: Narr.

Schubert, Klaus
2007 *Wissen, Sprache, Medium, Arbeit. Ein integratives Modell der ein- und mehrsprachigen Fachkommunikation.* (Forum für Fachsprachen-Forschung 76.) Tübingen: Narr.

Schubert, Klaus
2009 Positioning Translation in Technical Communication Studies. *Journal of Specialised Translation* 11: 17–30. http://www.jostrans.org/issue11/art_schubert.pdf (5. February 2009).

Shuttleworth, Mark and Cowie, Moira
[1997] 1999 *Dictionary of Translation Studies*. New ed. 1999. Manchester: St. Jerome.

Trubetzkoy, Nikolaj S.
[1939] 1989 *Grundzüge der Phonologie*. 7th ed. 1989. (Travaux du Cercle Linguistique de Prague 7.) Göttingen: Vandenhoeck & Rupprecht.

Whorf, Benjamin Lee
1940 Science and Linguistics. *Technology Review* 42 [6]: 229–231 + 247–248.

Wilss, Wolfram
1988 Übersetzen als Entscheidungsprozeß. In: Arntz, Reiner (ed.), *Textlinguistik und Fachsprache*, 7–20. (Studien zu Sprache und Technik 1.) Hildesheim/Zürich/New York: Olms.

Wilss, Wolfram
[1998] 2001 Decision Making in Translation. In: Baker, Mona with Malmkjær, Kirsten (eds.), 57–60. *Routledge Encyclopedia of Translation Studies*. New ed. 2001. London/New York: Routledge.

Colette Cortès

The class building potential of adjectives as used in German technical and advertising texts in a mail order catalogue

1 Introduction

Terminology research in languages for special purposes deals particularly with substantives while other word classes (verbs, adjectives) have been neglected for a long time. Nowadays, corpus-based studies allow a renewed approach for adjectives and verbs as terminologically relevant parts of speech. The subject I chose to write about is the use of adjectives to describe articles for sale in a mail order catalogue; it is a special kind of publication which develops two different intricated text types, one that is supposed to give the exact characteristics of the goods for sale on the one hand, and one that is supposed to persuade potential buyers of the eminent qualities of a product and of the necessity for them to buy it on the other hand.

The intention of this contribution is to show that in German for special purposes adjectives play a decisive role in the formation of compound terms coming from the observation that only a particular kind of adjectives can appear in a compound term of the form [adjective + N]. Furthermore, I will try to explain why German technical texts have developed a specific syntactic pattern, namely the structure [N + adjective] which is unknown in other text types. The terminologically relevant adjectives (adjective use) can be defined on the basis of a distinctive semantic-pragmatic property: its class building potential. This categorising function of some adjectives in a technical context is a semantic as well as a pragmatic feature, not of the adjective itself but of the adjective use, which is related to the technical or the advertising text respectively, as a frame conditioning the interpretation.

In this paper the syntactic behaviour of the adjectives (according to Milner 1978, see also Goes 1999 and L'Homme 2004) will be examined and the difference between the class building and non-class building use of adjectives will be defined (2); a corpus-based analysis will illustrate the class building potential of German adjectives (3), and we will finally focus on the relation between adjective use and text type (4). This work creates bridges between syntax, semantics, terminology and text linguistics.

To document the class building potential of German adjectives in a technical and an advertising text altogether, we will use a German mail order catalogue, the “Quelle-Katalog 2006”, that is to say the “Mode-Katalog” (MK) and the special issue for house equipment: “Ideen für ein schöneres Leben” (IsL), which were available until the end of August 2006.

2 About the class building use of German adjectives

To describe the class building potential of German adjectives, a complex syntactic-pragmatic operation has to be sketched on the basis of Jean-Claude Milner's work (1978). In his book published in 1978, *De la syntaxe à l'interprétation*, Jean-Claude Milner proposes a unified syntactic-semantic analysis of speech phenomena which are treated in different chapters in grammars and syntactic models like the interrogative and the exclamative sentence. Milner underlines the fact (which is true for many languages like French, German, English, etc.) that the interrogative and the exclamative sentences are often formed with the same morphemes (for example: *w*-morphemes + adjectives in German) and that, in spite of this morphological similarity, no ambiguity or interpretation problems are to be noticed, as examples (1) and (2) and the corresponding tests (1', 1") and (2', 2") show:

(1) *Da fühlte sich der Sänger ebenso als Künstler gewürdigt wie beim Geburtstagskonzert für eine 85-jährige Frau. Der Sohn hatte ihn auf der Straße gehört und spontan engagiert, verbunden mit dem Hinweis, dass die Mutter depressiv sei und seit Jahren kein Wort mehr gesprochen habe. Wiktor sang, und danach sagte die alte Frau zur Verblüffung aller Gäste einen Satz: “Was für eine schöne Stimme.”* (Der Spiegel 2000a: 104)

(1') *Ausrufesatz: Was für eine schöne Stimme! >> Wie schön ist seine Stimme!* (What a beautiful voice!/How beautiful is that voice!)

(1") *Fragesatz: ?Was für eine schöne Stimme hat er? // Was für eine Stimme hat er? Was für eine Stimmlage hat er = welche Art von Stimme hat er? eine Tenor-, Baryton-, Baßstimme usw.?* (What kind/type of voice has he?)

(2) *Dann geriet ein Holzpflock ins Sichtfeld der ferngesteuerten Kamera. In einer Tiefe von 95 Metern schwebte der Roboter über Sandstrände und überspülte Uferzonen hinweg. Der Lichtschein fiel auf versunkene Flussläufe und einen abgestorbenen Baumstamm, der wie ein totes Gerippe aus dem Meeresgrund ragte. Urwälder im Submarinen? Was für ein neo-*

lithisches Atlantis ist da in den Abgründen des Pontus Euxinus aufgetaucht? Nahezu vier Wochen lang wurde der Meeresgrund kartografiert. Dann war sich der wissenschaftliche Leiter der Expedition, Fredrik Hiebert, sicher: "Am Boden des Schwarzen Meeres haben einst Menschen gewohnt." (Der Spiegel 2000b: 267)

(2') *Fragesatz: Was für ein neolithisches Atlantis (= eine Art von neolithischem Atlantis) ist da in den Abgründen des Pontus Euxinus aufgetaucht? (What kind of Neolithic Atlantis appeared in the abysses of Pontus Euxinus?)*

(2") *Ausrufesatz: ? Was für ein neolithisches Atlantis!/*Wie neolithisch ist dieses Atlantis!/*Wie neolithisch dieses Atlantis ist! (What a Neolithic Atlantis!/*How Neolithic is this Atlantis!/*How Neolithic this Atlantis is!)*

We can sum up Milner's results which are relevant for the German language in the following way: The scope of an interrogative sentence is not the same as the scope of an exclamative sentence, and the introducing *w*-expression *was für ein* is therefore a good revealing point:

- With *was für ein N* in an interrogative sentence, *N* has to be understood as a superordinate and the speaker asking the question expects a statement about a subordinate, as proves the possible paraphrase with: *welche Art von?* (what kind of?) in (1") and (2'), a paraphrase which would be impossible in the exclamative sentence, or if needed just with an ironic overtone (**Was für eine Art von schöner Stimme!* *What a kind of beautiful voice!). The scope of the interrogative sentence introduced by *was für ein N* is *N* as a hypernym. Crucial for the interpretation of the interrogative sentence introduced by *was für ein N* is a class building semantic operation and the corresponding membership relation.
- The exclamative sentence W*hat a ADJ N!/How ADJ is that N!* (*how* + adjective!) has a quite different function: it expresses the speaker's subjective point of view, and not an objective classification. With *was für ein N* in the exclamative sentence, an evaluative scale is opened depending on criteria that refer to the judgment of the speaker. The exclamative sentence appears only in a subjective text, that is to say a text which has to be related to the presence of the speaker in his discourse, imposing a specific interpretation based on a necessarily not objective viewpoint. (Kerbrat-Orecchioni 1980)

This clear differentiation between the world-knowledge-related interrogative sentence and the speaker-related exclamative sentence is responsible for the fact that in each type of sentences only some kinds of adjectives can be used. Consider the adjectives contained in examples (1) and (2) *schön* and *neolithisch*: only *schön* and not *neolithisch* can be used in an evaluative, speaker-related context like *ich finde X* adjective: *Seine Stimme finde ich schön* ≠ **Die versteinerten Bäume von Chemnitz finde ich neolithisch.* Hence there are adjectives like *schön* in (1) which are compatible with the exclamative sentence, and others like *neolithisch* in (2) which are compatible with the interrogative sentence. Milner speaks of *distribution complémentaire* (complementary distribution) to characterise the relation between these two differentiated uses of adjectives. The adjectives which can be used only in interrogative sentences bear the feature [+ class building] (*classifiants*) while the adjectives which can be used only in exclamative sentences bear the feature [– class building] (*non classifiants*). The property [+/– class building] results from the syntactic behaviour of the adjectives, as shown by a series of tests proposed by Milner 1978 and applied to the German language:

- the predicative use of adjectives,
- the negative use of adjectives,
- the anaphoric use of adjectives. (See Table 1)

	class building adjectives *rot* and *blau* (3)	**Non-class building adjectives *herrlich* (4)**
Predicative use	*Dieser Punkt ist blau. Der da ist ein blauer Punkt.* Judgment about the class-membership of a point.	*Das Schloss ist herrlich, finde ich herrlich*: No class building operation but speakers's subjective evaluation.
Negative use: negation + self-correction (*nicht … sondern*)	*Dieser Punkt/Der da ist nicht rot, sondern blau.* Negation + correction: changes the class-membership of the point.	*Das ist nicht herrlich, nicht hervorragend, nicht wunderbar*: No class-membership change, but negative evaluation.
Anaphoric use	*Die Roten verschwenden ihre Energie (…). Die bescheidenen Blauen gewinnen.*	**Meine Lieblingsschlösser sind die Herrlichen.* Anaphoric use not allowed.

Table 1: Class building and non-class building adjectives

Another specific test typical of the German language has to be added, that is the possibility to use the adjective as a part of a compound term [adjective + N]. Form- or colour-adjectives like *rund, oval, blau, grün, rot* are class building adjectives, and high degree qualifying adjectives like *herrlich, hervorragend, prima, wunderbar, schrecklich* are non-class building ones. Table 1 illustrates the two different kinds of interpretations that those adjectives allow. With a class building adjective (*rote und blaue Pünktchen*) in a predicative sentence, a judgment is delivered about the class-membership of a "little point" while the use of a non-class building adjective (*die herrlichen Schlösser Neuschwanstein und Hohenschwangau*) expresses a subjective evaluation (examples in Table 1 are extracted from (3) and (4) below).

(3) *Optisch gleicht sein Modell einem Computerspiel: Auf dem Bildschirm wandern rote und blaue Pünktchen umher und tilgen grüne Felder. Wenn sie genug gefressen haben, vermehren sie sich. Wie bei Militärmanövern sind die Blauen die Guten: Sie ernähren sich langsamer und geben den grünen Feldern Zeit, wieder nachzuwachsen. So haben auch ihre Nachkommen etwas davon. Die Roten dagegen machen alles platt, vermehren sich dafür aber wie wild. Peppers Ergebnis: Wenn genügend Felder schön dicht beieinander stehen, lohnt sich der Egoismus. Nach wenigen Minuten wimmelt der Bildschirm vor roten Punkten; die blauen sind ausgestorben. Liegen die Felder dagegen weit auseinander, verschwenden die Roten ihre Energie, indem sie hungrig im selbst verschuldeten Kahlschlag umherirren. Die bescheidenen Blauen gewinnen.* (Der Spiegel 2000c: 156)

(4) *"Ludwig II. – Sehnsucht nach dem Paradies" ist ein neues Glanzstück im bayerischen Kulturleben, eine Huldigung an den bayerischen Märchenkönig in Musik und Optik, Klangkompositionen und Bühnenbild. In dramatischen Szenen werden die Visionen des Königs zwischen Sein und Schein theatralische Wirklichkeit. Das "Ludwig"-Theater ist die weltweit erste Musicalbühne am historischen Originalschauplatz, gebaut am Forggensee bei Füssen, mitten hinein in den bayerischen Königswinkel mit Blick auf die weltberühmten, herrlichen Schlösser Neuschwanstein und Hohenschwangau.* (Der Spiegel 2000d: 118–119)

As shown in Table 1, the use of a class building adjective allows the differentiation of a subclass (*die blauen Punkte: the blue points*) within a broader class (*die Punkte: the points*). Evidence for that is provided by a series of tests developed by Milner (1978):

– in a sentence with the class building adjective *blau* used in a predicative way (*Dieser Punkt ist blau*) a membership relation is assumed; in other words, the point belongs to the class comprised of all blue points. In that way, the class of the blue points is opposed to the class containing the red points for example. On the contrary, a non-class building adjective in a predicative sentence has to be interpreted as a subjective opinion of

the speaker (**Das finde ich blau/*I find it blue*)/(*Das finde ich herrlich/I find it wonderful*).
- the negative and corrective expression *nicht ... sondern ...* (*not ..., but ..*) in a sentence with the class building adjective *blau* used in a predicative way articles a (sub)class-change while the negation of a sentence with a non-class building adjective used in a predicative way has another effect, namely the transition from a positive (*wonderful*) to a negative evaluation (*not wonderful = bad, awful, ugly*).
- the anaphoric use of class building adjectives, which can be considered as a kind of pronominalisation, is a usual and productive syntactic and lexical process (see *die Roten, die Grünen/the reds, the greens*) and so on in many different fields (politics, sport, and so on), which is typical of terminology and which is based on the use of class building adjectives. On the contrary, the pronominalisation of non-class building adjectives (**die Herrlichen, *die Schrecklichen/*The wonderful, *the awful*) creates no productive pattern.

The use of a class building adjective refers to a taxonomic relation drawn on Figure 1. *Punkte* functions as the hypernym and *die blauen Punkte* and *die*

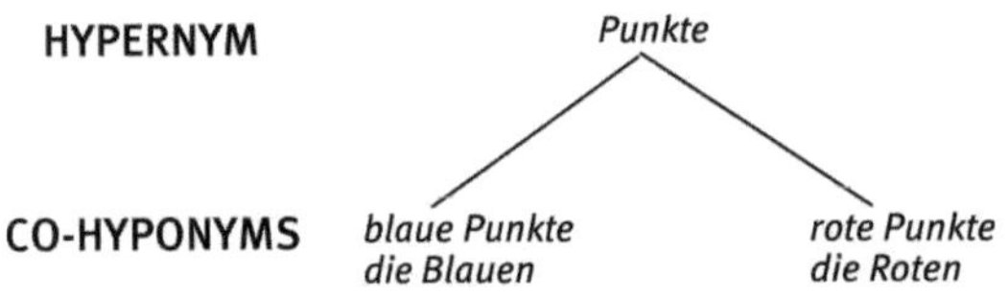

Figure 1: The use of a class building adjective refers to a taxonomic relation

roten Punkte are cohyponyms. As a matter of fact, class building adjectives are subclass building adjectives which mark the relation between the generic substantive (*Punkte*) and the noun phrases with specifying class building adjectives (*die blauen Punkte* and *die roten Punkte*) in an efficient and economic way.

It must be underlined that the opposition between class building and non-class building adjectives differentiates only partly separated lists of adjectives because many adjectives can have both a class building and a non-class building use. The adjective *schön* (*beautiful, fine*) for example, which is not used as a class building adjective in example ((1) *Was für eine schöne Stimme!/How beautiful that voice is!*), can have a class building function in a term like *die Schönen Künste* (*les Beaux-arts/the Fine Arts*). From this observation we can draw three kinds of conclusion:

- Besides class building adjectives like *neolithisch* and non-class building like *herrlich*, there is a large two-part-class of adjectives like *schön*, which can perform both functions according to the context. Hence the feature [+/– class building] does characterise not the adjective itself but the adjective use.
- The expression *die Schönen Künste* can be analysed as a compound term; in our corpus we have to examine the relation between class building adjective use and nominal composition.
- As Figure 1 shows, the class building use of an adjective in a noun phrase [Det + Adj + N] presupposes the existence of a paradigm of possible cohyponyms (*die Schönen Künste* (*Beaux-arts/Arts*) ≠ *die bildenden Künste* (*arts plastiques*) ≠ *die sieben Freien Künste* (*arts libéraux*)). The existence of a paradigm of cohyponyms is a new relevant criterion that provides evidence for the class building use of an adjective, and that must be taken into account as a relevant test that must be applied to our corpus data.

The opposition between class building and non-class building adjectives can be documented very precisely in a mail order catalogue like the "Quelle-Katalog".

3 Analysis of the feature [+/– class building] in the 2006 Quelle-Katalog

Typical of technical texts is the fact that besides the combination [inflected adjective + N] the structure [N + non-inflected adjective] is often encountered, and we would like to show that the feature [+ class building] offers a convincing explanation for both phenomena. We begin the analysis of our corpus data extracted from the 2006 Quelle-Katalog with class building adjectives (firstly the preposed (3.1.1) and secondly the postposed class building adjectives (3.1.2)), and we will pursue with non-class building adjectives (3.2).

3.1 Class building adjective use in the 2006 Quelle-Katalog

3.1.1 [Adjective + N]: preposed, inflected class building adjectives

Let us examine an example on the basis of a series of tests registered in Table 2.

(5) (IsL) page 148: *mechanische Waage ≠ Digitalwaage ≠ (Körper-)Analysewaage (Körperfett- und Körperwassermessung auf 0,1 % genau.)*

Interrogative sentence: possible	*Was für eine Waage möchten Sie kaufen? eine mechanische Waage oder eine Digitalwaage oder eine Analysewaage?*
Exclamative sentence: impossible	*Was für eine mechanische Waage!/Wie mechanisch ist diese Waage!/Wie mechanisch diese Waage ist!*
Predicative use of the adjective: unusual	*Dies ist eine mechanische Waage/Diese Waage ist mechanisch.* Better: *Diese Waage ist eine mechanische Waage.*
Negative use of the adjective (+ *sondern*: possible)	*Dies ist keine mechanische Waage, sondern eine Digitalwaage*
Anaphoric use of the adjective: possible	*Von all diesen Waagen gefällt mir die mechanische am besten.*
Existence of a paradigm of cohyponyms: possible	*mechanische Waage ≠ Digitalwaage ≠ (Körper-) Analysewaage*
Paradigm [inflected adjective + N] ≠ compound noun: possible	*mechanische Waage ≠ Digitalwaage ≠ (Körper-) Analysewaage*

Table 2: The results of the series of tests

The results of the series of tests consigned in Table 2 show that the expression *mechanische Waage* is a technical term belonging to a paradigm of different pairs of scales. As regards the form, only *mechanische Waage* is constituted according to the pattern [inflected adjective + N]; the other co-hyponyms correspond to the patterns [non-inflected adjective + N] (*Digitalwaage*) or [N + N] (*Analysewaage*). However, the semantic function of the determining preposed constituent of the technical term remains the same: it is always [+ class building].

On the syntactic level, the adjective *mechanisch* or *digital* has a syntactically restrictive and a semantically distinctive function; it serves to delimit a subclass in a kind of subcategorising process.

The morphological criteria plays only a secondary role and many possible word classes can perform this distinctive function like:

- Participles: (IsL 198: *selbstklebende Dekorfolie/selbstklebende Spiegelfolie*) (IsL 229: *wärmedämmende Deckenplatten*) (IsL 382: *(Navigationssysteme) motorisierter T3 Monitor mit exzellenter Bildqualität und großem Betrachtungswinkel).*
- Compound adjectives: (IsL 390 (5): *Das Design-Duo: 1 schnurloses + 1 schnurgebundenes Telefon.*).
- Relational, derivative adjectives (IsL 42 (1)/46 (1): *3-teilige Eckgarnitur*) (IsL 42 (3): *3-sitziges Sofa*) (IsL 305: *Integrierbare Einbau-Geschirrspüler*).

All class building adjectives have something in common: they are not gradable, precisely because of their distinctive function. That on the contrary adjectives with [– class building] use cannot be encountered in interrogative sentences proceeds from the fact that a question cannot focus on the subjective appreciation relation implied in the gradability while the question whether an object belongs to a class or not can provide the focus of an interrogative sentence which allows a distinctive yes- or no-answer. The examined class building adjectives and participles are used to name a subclass of N according to the [distinctive determiner + N] pattern; this nomination process is based on taxonomic relations.

3.1.2 [N + adjective]: postposed, non-inflected class building adjectives

We are trying here to provide an account of a fact that is neglected by even the best grammars of the German language: the fact that the [N + adjective] pattern with postposed, non-inflected adjectives (that is to say: without morphological agreement) can be found in a German mail order catalogue.

The Duden-Grammatik (2005: §467) only notices: "Nachgestellte unflektierte Adjektive erscheinen öfter bei Produktbezeichnungen in der Fachsprache: Henkell trocken, Werkzeugsschränke grün RAL 6011." For the IdS-Grammatik (Zifonun, Hoffmann and Strecker 1997, 1991–1992) the [N + adjective] pattern with postposed, non-inflected adjective cannot be analysed: While an expression like *Forelle blau* can be dealt with as an "Adverbialkonstruktion in elliptischer Verwendung z.B. auf Speisekarten", "liegt wieder etwas anders der Fall Henkell trocken, Coca Cola eisgekühlt, wo wohl ursprünglich bzw. als Konstruktionsmuster (elliptische) Adverbialkonstruktionen oder nachgestellte Adjektive vorliegen. Diese sind aber in ihrem aktuellen Gebrauch zu Marken- bzw. Gattungsnamen verfestigt und als solche nicht mehr syntaktisch analysierbar".

On the basis of the class building potential of German adjectives and of the distinctive nomination process which they allow, we would like to propose a tentative explanation for the use of postposed, non-inflected adjectives in a German mail order catalogue.

Let us begin with adjectives which are always class building ones. Many colour adjectives are used (see (6) below) as postposed and non-inflected adjectives, although their function as class building adjectives is not modified. The special [N + adjective] pattern just underlines the fact that the colour in (6) has not a describing function but that it has to be interpreted exclusively as a distinctive feature. The same is true about the form of the glass cabinet in example (7).

(6) IsL 87 (2/3): *Matratze blau ≠ Matratze rot.*

(7) IsL 53 (6): *Vitrine 1-türig, mit Beleuchtung.*

However, the true justification for the use of postposed, non-inflected adjectives can be found by adjectives which happen to be used in both a class building and a non-class building context. The postposition with non-inflectedness works in that case as a desambiguiser: the adjective must be interpreted then as a distinctive and hence not gradable class building marker. That is particularly important for adjective pairs like *groß* and *klein* or *breit* and *schmal*, which allow very subjective interpretations when they are not quantified (*1 Meter groß, 2 Meter breit*).

(8) IsL 52 (4): *Bank in 2 Größen: Bank klein ≠ Bank groß*

(9) IsL 62 (2): *Anrichte breit ≠ Anrichte schmal*

When they are postposed and non-inflected as in (8) and (9), the adjectives *groß* and *klein* or *breit* and *schmal* perform the distinctive nomination function of exclusively class building adjectives.

3.2 Non-class building adjective use in the 2006 Quelle-Katalog

The non-class building use of adjectives cannot perform a distinctive nomination function; they appear in exclamative sentences introduced by *was für ein* or *wie*, where the adjectives are used predicatively; they can be pronominalized only when they have the superlative form. The complex pattern [non-class building adjective + N] is not a term: it cannot appear in a series of cohyponyms. (See Table 3)

(10) IsL 99: *Die besondere Qualität. Dieser hochwertige Stoff aus 100 % Baumwolle ist hautfreundlich and dabei besonders strapazierfähig.*

The gradability of the adjectives in the Quelle-Katalog is expressed by different means: a participle in (11: *strahlend schön*), an adverb in (13: *vielseitig kombinierbar*) or a compound adjective in (14: *Topaktuell*).

(11) IsL 236: *Die neue Eleganz für ein strahlend schönes Zuhause.*

Interrogative sentence: impossible	**Was für einen hautfreundlichen und strapazierfähigen Stoff möchten Sie kaufen?*
Exclamative sentence: possible	*Was für ein hautfreundlicher und strapazierfähiger Stoff!/Wie hautfreundlich und strapazierfähig dieser Stoff ist!*
Predicative use of the adjective: possible	*Dieser hochwertige Stoff aus 100 % Baumwolle ist hautfreundlich und dabei besonders strapazierfähig.*
Negative use of the adjective (+ sondern: impossible)	*Dieser Baumwoll-Stoff ist nicht besonders hautfreundlich (Negative Einschätzung), *sondern ...*
Anaphoric use of the adjective: impossible	*Von all diesen Stoffresten gefallen mir die hautfreundlichen und strapazierfähigen am besten.*
Existence of a paradigm of cohyponyms: impossible	*hautfreundlicher und strapazierfähiger Stoff ≠ ?*
Paradigm [inflected adjective + N]/≠ compound N: impossible	*hautfreundlicher und strapazierfähiger Stoff ≠ ?*

Table 3: Non-class building adjective use

The gradability seems at first sight to be the decisive feature for the non-class building use of the adjectives. Paradoxically, there are, however, non-class building adjectives which cannot be graded like *ausgezeichnet* oder *herrlich*, but which cannot be said non-gradable for that reason; they refer to the end level of an evaluation scale and can be called: “high-grade” adjectives. We must then give up strict gradability as the decisive feature for the non-class building use of adjectives, and the existence of a subjective evaluation scale can be considered as the main criterion to define the semantic properties of non-class building adjectives. An evaluation scale presupposes an evaluating speaker and that is the reason why non-class building adjectives appear in a subjective text.

Our last chapter will be dedicated to the correlation between the adjective use and the definition of the corresponding text type.

4 Text typology and syntax of class building adjectives in the 2006 Quelle-Katalog

We will consider the Quelle-Katalog as a cursive text, although most of the descriptions are in telegraphese style, that is to say without phrase-binding and partly without morphological inflection (like for example: *Shirt Details:*

runder Ausschnitt. 1/2 Ärmel, Länge ca 58 cm.) The text of a mail order catalogue can be considered as a mix of two intricated text types, an advertising text addressing potential buyers in a pleasant, flattering way, and a descriptive, objective technical text, which functions as a contract between the mail order company and the customers.

We will demonstrate the existence of both of these text types with some examples from the 2006 Quelle mail order catalogue. On page 30 of the Mode-Katalog (MK), the presentation of the goods follow a few simple structuring guidelines, and this structure is applied repeatedly in the whole catalogue with only small variations. We can see on the side of the picture a description that we call "picture description". The text description itself consists of three parts (see example (12)):

- a first text section which brings arguments for buying the described article (*Shirt: Der fantasievolle Muster-Mix aus Blüten und Schriftzug ist zusätzlich mit glitzernden Swarowski-Steinchen versehen – ein trendstarkes Basic.*),
- an objective description entitled: *Details: runder Ausschnitt. 1/2 Ärmel, Länge ca 58 cm.*
- and another objective description entitled: *Material: 100 % Baumwolle, gekämmt. Maschinenwäsche.*

The picture description and the first part of the text description contain partly complete sentences addressing the customer with *Sie* (which is the addressee-pronoun used by a speaker who observes the formalities of polite society), and many adjectives with non-class building use. The objective descriptions entitled: *Details* or *Material* are worded in telegraphese style and exclusively contain adjectives with class building use. The picture description and the first part of the text description can be considered as a subjective customer friendly advertising text and the *Details* or *Material* descriptions as an objective technical text containing the contractual obligations which have to be honoured by the mail order company towards the customers.

Typical of the advertising text addressing the buyer are comments like:

- (12) (MK) page 30 (2) *Shirt: (...) – ein trendstarkes Basic.*
- (13) (MK) page 30 (3): *Absolut im TrendBlazer in attraktiver Nappa-Optik. Feminin taillierter Style – von elegant bis lässig unglaublich vielseitig kombinierbar.*
- (14) (MK) page 30 (4): *Tasche. Topaktuell.*

This short list shows how important the use of valuation adjectives can be for advertising texts. In addition to this can be observed subjective gradation markers, like *unglaublich* in *unglaublich vielseitig kombinierbar* (13), which would not appear in an objective technical text.

(12) (MK) page 30 (2): *Baumwoll-Shirt mit Swarowski-Steinchen*
Shirt: Der fantasievolle Muster-Mix aus Blüten und Schriftzug ist zusätzlich mit glitzernden Swarowski-Steinchen versehen – ein trendstarkes Basic.
Details: runder Ausschnitt. 1/2 Ärmel, Länge ca. 58 cm.
Material: 100 % Baumwolle, gekämmt. Maschinenwäsche

(13) (MK) page 30 (3): *Absolut im Trend: Blazer aus weichem Porc-Velours, in Nappa-Optik. Echtes Leder. Blazer in attraktiver Nappa-Optik.*
Feminin taillierter Style – von elegant bis lässig unglaublich vielseitig kombinierbar.
Details: Revers-Kragen, mit 2 Knöpfen zu schließen. 2 Taschen mit Klappen. Länge ca. 60 cm.
Material: Porc-Spalt-Velours, Futter 100 % Polyester. Lederreinigung.

(14) (MK) page 30 (4): *Tasche in Veloursleder-Optik*
Tasche. Topaktuell. in Beutelform und Veloursleder-Optik
Details: großes Hauptfach mit Magnetverschluss und Zugband. Handyfach. Innen-, Vor- und Steckfach jeweils mit Reißverschluss. Verstellbarer Schulterriemen ca. 31 x 35 x 16 cm.
Material: Synthetik

There are also on some pages longer advertising texts with complete sentences, especially exclamative sentences: *Die besondere Qualität. Gibt Flecken keine Chancen! Machen Sie sich das Leben ein bisschen leichter!* ((15) (MK) page 193 (4)) or *Nehmen Sie es leicht, mit Jersey! Ideal auf Reisen! Angenehm auf der Haut* ((16) (MK) page 195 (4)). Gradation markers can be found there too: *ein bisschen leichter, schön pflegeleicht*, in addition to non-class building, subjective, valuation adjectives: *angenehm, perfekt, ideal*, or ((MK) page 246): *Silberschmuck von s. Oliver, schick und trendy!* Subjective, valuation nouns are also used in advertising texts: *der besondere Clou.* ("high-grade" substantive).

(15) (MK) page 193 (4):
Die besondere Qualität. Gibt Flecken keine Chancen! Machen Sie sich das Leben ein bisschen leichter: diese sommerliche Hose aus weicher Baumwolle kühlt angenehm. Dabei sorgt der Elasthan-Anteil für völlige Bewegungsfreiheit. Der besondere Clou: Die Oberfläche wirkt Schmutz abweisend. Easy life eben!

Hose mit Dehnbund in 4 Farben (weiß, Sand, kiwi, marine). Mit Easy-life-Ausrüstung.
Hose. Die sitzt perfekt und gibt Ihnen völlige Bewegungsfreiheit, dafür sorgt die hautfreundliche Baumwoll-Stretch-Qualität! Die Behandlung der Faser im Easy-life-Verfahren bringt Ihnen noch zusätzliche Vorteile: die Oberfläche wirkt Schmutz abweisend. Dabei ist diese Hose extrem pflegeleicht und verfügt über lange Waschbeständigkeit.
Weitere Details: seitlicher Dehnbund. 2 Taschen vorn. 2 Gesäßtaschen mit dekorativer Biesenverarbeitung. Länge: N-Größe ca 104 cm, K-Größe ca 99 cm.
Material: 98 % Baumwolle, 2 % Elasthan, Maschinenwäsche.

(16) (MK) page 195 (4): *Nehmen Sie es leicht, mit Jersey! Ideal auf Reisen: glatt ohne Bügeln*
Blusen aus weichem Jersey. Angenehm auf der Haut und schön pflegeleicht.
Details: Messingknöpfe, 2 Brusttaschen, leicht tailliert, kleine Seitenschlitze. 3/4 Ärmel mit Schlitz
Material: 50 % Polyester, 50 % Viskose. Maschinenwäsche.

Although some explanations or definitions like *Zirkonia (Diamant-Imitat)* in (17) (MK) page 246 are admittedly customer friendly, they must, on the basis of their knowledge transferring objectivity, be considered as parts of the technical text.

(17) (MK) page 246: *Silberschmuck von s. Oliver, schick und trendy!*
(2) *Anhänger mit Kette, 28 Zirkonia (Diamant-Imitat) Länge 43 cm.*
(4) *Ensemble mit farbigen Perlen (synth.) und Silberkugeln. Armband. Länge 19 cm.*

With the help of examples (12) to (17) were defined the main characteristics of the advertising text, a text structure addressing the customer, with exclamative sentences, gradation markers and subjective non-class building adjectives.

Typical of the descriptive technical text are simple assertions, generally worded in telegraphese style, no interrogative and no exclamative sentences, no gradation markers and only class building adjectives (*runder* Ausschnitt/Baumwolle, *gekämmt/weiches* Porc-Velours/*großes* Hauptfach/*verstellbarer* Schulterriemen/*seitlicher* Dehnbund/*farbige* Perlen). The descriptive technical text concerning every article consists of the three following parts:

- The entry, which has a categorising function for the concerned article (*Shirt, Blazer, Tasche, Hose, Schmuck*), comprises at least one substantive.

There are generally no adjectives on that level, and if any, then only class building ones (*elektrische Zahnbürste/electrical tooth-brush*).

- The following remarks belong to the advertising text, from which specific descriptive elements can be extracted: *Blusen aus* (...) *Jersey*. The picture description, which belongs generally to the advertising text, can also have a specifying effect and can then be considered as a part of the technical text like *Baumwoll-Shirt mit Swarowski-Steinchen.*
- Under the rubrics: *Details* and *Material*, we find other specific features which belong to the technical text.

In the Quelle-Katalog the typical technical text serves to describe an article (or a type of article) with class-denomination (generally with a mere generic

Description of an article [Class-denomination generic substantive + specific characteristics

Figure 2: Description of an article

substantive) and specific characteristics (see Figure 2). Even if the customer finally buys a specimen of the article, the purpose of the descriptive technical text of the Quelle-Katalog is not the description of an object, but the description of an object type; with the delimitation of generic classes and specified subclasses, it tends to a true article ontology, wherein the adjective plays a decisive role: a mere adjective is sufficient to word the objective, class building intention (*elektrische Zahnbürste*) or the subjective publicity intention of the message (for example *Ideal auf Reisen* (...). *Angenehm auf der Haut*. usw.).

The telegraphese style used in the technical text is not chosen only to save place and paper; it must be related to the purpose of the speaker: it characterises the technical text and can be explained with the tentative construction of an article ontology. What is described is not the object that is finally sent to the buyer, but an article type that could be identified with the corresponding number in the computer program of the Quelle mail order company. The article type need not be referred to as a definiteness-marked noun phrase: *Tasche, in Beutelform und Veloursleder-Optik, großes Hauptfach mit Magnetverschluss und Zugband. Handyfach. Innen-, Vor- und Steckfach jeweils mit Reißverschluss. Verstellbarer Schulterriemen ca 31 x 35 x 16 cm.* The determiner (article, possessive, demonstrative, etc.) could appear only in an actualised sentence referring to the situation of speech: *Die/Ihre Tasche in Beutelform und Veloursleder-Optik Nummer 50 18 651 bringt Ihnen der Postbote direkt ins Haus.* Such a sentence could appear only in an advertising text. In a technical context the description situation in the mail order catalogue should be referred to: *die hier beschriebene Tasche hat ein Handyfach, ein Innen-, ein Vor- und ein Steckfach* (...), an expression that must be interpreted as the description of an object and not of a type. The technical text in the Quelle-Katalog, which is based on ontology, is not related to the situation of speech, but to the world-knowledge; therefore the telegraphese style is the most suitable style to put into words the ontological construction and to release the class building operation from the situation of speech. This offers also an explanation on the text level for the fact that a class building adjective used in such a context can be postposed and then cannot be inflected (see 3.1.2).

In the Quelle-Katalog, both text types (subjective, speaker-related advertising text and objective, world-knowledge-ontology-related technical text) are associated, so that besides the dry, unemotional description of the article (with an objective technical text type, class building interpretation of the adjectives and constitution of an ontology of the sold goods) appears a subjective advertising text with non-class building, qualifying adjectives, which strongly advises [subject = the best] the reader to buy the described goods. Besides the "cold-hearted" class building technical text, often worded in telegraphese style, stays the "warm-hearted", customer-friendly advertising text, which plays the role of a friendly salesman trying to persuade the potential buyer of the excellence of the goods and of the absolute necessity to purchase them.

5 Summary and results

The [+/– class building] feature is very important to define the function of adjectives, but it can achieve different effects on many other levels of the linguistic construction:

- on the level of the word-composition,
- on the level of the syntax of the noun phrase,
- on the level of the sentence structure,
- on the level of the text structure.

Considering the extent of the [+/– class building] feature and its numerous applications, it can be assumed that it is part of the speaker's competence:

- On the level of the word composition: The determining, preposed element X of a compound term [X + N] used to name a subclass of N is [+ class building] whether X is a substantive, an adjective or a preposition.

- On the level of the syntax of the noun phrase: An adjective is [+ class building] when it is used in a compound term; a class building complex [adjective + N] can be used without a determiner if it refers to a world-knowledge-related ontology and not as a constituent of a sentence which is per definitionem related to the situation of speech.
- On the level of the sentence structure: [+ class building] (non gradable) adjectives are used in interrogative sentences while [– class building] (gradable or high degree) adjectives are used in exclamative sentences.
- On the level of the text structure: This issue provided evidence for the fact that the Quelle-Katalog contains two opposite text types which can be distinguished on the basis of the class building potential of the German adjectives: a technical text with [+class building] adjectives and an advertising text with [– class building] adjectives.

In short, the [+/– class building] feature is a semantic marker with effects on the lexicon and on the syntactic and pragmatic structure of sentences and texts.

The relevance of such a feature for the German language makes the formalisation of syntactic rules difficult because it precisely questions the autonomy of a syntactic component in a linguistic model just as Milner (1978) already assumed. But its main benefit is to bridge the gap between domains of the linguistic research like terminology and syntax which are scarcely connected to each other, and to open the way to promising collaborations.

References

Der Spiegel
2000a *Der Pavarotti von der Alster*, von Schrep, Bruno. No. 43, p. 104.

Der Spiegel
2000b *Strafgericht am Bosperus*, von Schulz, Matthias und Zand, Bernhard. No. 50, p. 267.

Der Spiegel
2000c *Die Grenzen der Erkenntnis. Rätsel des Lebens. Gertrude – gerettet! Komplexität*, von von Bredow, Rafaela. No. 50, p. 156.

Der Spiegel
2000d *Ludwig II. – Sehnsucht nach dem Paradies*. No. 53, p. 118–119.

Duden
2005 *Die Grammatik*. (Duden Band 4.) 7. völlig neu erarbeitete und erweiterte Auflage. Mannheim: Bibliographisches Institut & E.A. Brockhaus AG.

Goes, Jan
1999 *L'adjectif. Entre nom et verbe*. Paris/Bruxelles: De Boeck & Larcier, Département Duculot Champs Linguistiques.

Kerbrat-Orecchioni, C.
1980 *L'énonciation. De la subjectivité dans le langage*. Paris: Linguistique, Armand Colin.
L'Homme, M.-C.
2004 Adjectifs dérivés sémantiques (ADS) dans la structuration des terminologies. Conférence "Terminologie, ontologie et représentation des connaissances" organisée à l'Université Jean Moulin Lyon I, Atala 22 et 23 janvier 2004.
Milner, Jean-Claude
1978 De la syntaxe à l'interprétation. Quantités, insultes, exclamations. In: *Travaux*, Paris: Linguistiques, Seuil.
Quelle-Katalog
2006 *Mode Katalog (Fashion, Summer) and Ideen für ein schöneres Leben (House Equipment)*.
Zifonun, Gisela, Hoffmann, Ludger and Strecker, Bruno
1997 *Grammatik der deutschen Sprache*. 3. Band. (Schriften des Instituts für Deutsche Sprache 7.3.) Berlin/New York: de Gruyter.

Clemens Schwender

Instructional illustrations: Waypoints in the historical development

1 Introduction

We began making pictures, the evidence suggests, more than 35,000 years ago. Only a few pictures from this period have been preserved, in caves, and include the occasional image of humans at work: hunting and fighting. The images also show humans using technology: we see figures using tools and weapons. It might therefore be worthwhile to take a look at these historical images, which, without any obvious purpose, demonstrate humans dealing with technology. Historians disagree about the significance of these early drawings, but nonetheless, they still are a part of the history of the depiction and use of technology; this is where the early development of our pictorial skills began. Still, the question remains: Can they serve as a point of departure for an analysis of the visualization of technology?

It is not possible to date the beginning of the use of technical illustrations for instructional purposes. Technical documentation can be defined as a written and/or pictorial illustration of how to build or use an artefact. The target groups are laypersons. We must therefore look for those exemplary visual building blocks that, having arisen for the most varied of purposes in differing contexts and at differing times, have gained general acceptance for instructional purposes. Time and again designers have made use of developments intended to solve other problems, but whose elegant solutions have also served to solve problems in depicting technology. One must also consider illustrations that demonstrate actions, as well as the occasional use of technology, or offer specific aspects that could contribute to a history of the use of images in instructional materials. There only remains the question of just what instructional illustrations are and how to identify them.

Pictures can depict, i.e., they offer us a visual reproduction (leaving aside considerations of abstract art for the time being). They can show us our thoughts and imagination: religious and cultural beliefs and the afterlife, for example. When drawn from the imagination or remembrances of the painter, they can illustrate the idealized; the painter can choose to paint a house or this house. The picture may be derived from fantasy or from memory. It can also be based upon a pattern that the draftsperson or painter follows. Even more varied than its representational relationship to its model is its intent. What did the artist have in mind? Was it painted out of boredom? Was it an effort at personal expression, intended to make an aesthetic statement? Or was it intended to please a patron? Are we dealing with a literal representation – a concrete embodiment – or the idea or concept that it represents? Has a specific moment in time been captured? Is it intended to elicit from the observer specific thoughts, associations, feelings, or even actions? Pictures are capable not only of showing us how things are, but also how they should be. When we then add to this intent – a direction to the observer to carry out a specific task in order to arrive at a specific state – such illustrations become a part of our investigation. When these conditions are met, we can speak of instructional illustrations, pictures explaining an activity or action. In the following examples not all of these conditions will be fully met; nonetheless we are interested in following the trail of those stylistic conventions that have made their way into the history of visual instruction. Before undertaking such a search, we must first pose another question: Why do we make images? This is not a question of the motivation behind a specific image, but rather of psychological capabilities.

Humans are not only capable of following a fixed stimulus-response pattern, but also of planning at least some aspects of their activities and behavior; with free will there is some latitude. In order to balance out the disadvantages of time-consuming thought and deliberation, we are able to mentally anticipate our actions. This visual anticipation occurs in the visual cortex, a portion of the brain in the rear of the head. The association areas also play an important role, allowing us to construct a model of the world in our mind and manipulate, plan, and consider various possibilities. In situations allowing an idea or plan, various modes of behavior are tested and tried out. According to Aebli (1981: 310), thought reveals itself as a cognitive activity in the "planning, executing, and judging of actions, and of factual and social processes and situations that influence actions." These thoughts or ideas can be recorded either verbally (written or otherwise makes no difference) or by means of images (still or moving). They thus remain preserved, recallable, and communicable at other times and places. One can look at these with respect to their communicative function as trial actions, in historically developed forms, genres, and styles, that view, explore, and socially construct reality as possibility.

The option of recording factual situations brings a further significant advantage: we no longer have to keep everything in our head, thus relieving our memory and expanding its capacity. This insight is by no means new. It was already noted in a book on fireworks displays dating from 1420: "And that is why, because the pieces comprising it are so numerous which any decent armaments master should be capable of and which a master without writings cannot keep in his mind, that is why everything useful and necessary is written down hereafter" (Hassenstein 1941: 43).

The chronicle therefore serves as a memory aid, since it is impossible to remember everything one is supposed to know. As early as A.D. 97, Sextus Julius Frontinus wrote in his book *De Aquaeductibus* about plan drawings:

> We also had plans of the ducts drawn up, from which can be seen where the valleys and rivers lay that were bridged over, how large they are, and where precise and ongoing monitoring and maintenance of the hollowed channels built across mountain slopes is necessary. This has the advantage that we have the objects right before our eyes and can refer to them as if we were standing right in front of them (Feldhaus 1959: 11).

Thus we have mentioned the most important prerequisites for the development of instructional materials and illustrations: We can medially record that which is conceivable and thereby expand our memory, and we can communicate these manifest contents. Text and image aid the explanation or clarification. That which is not current is thus updated. Words and images are made visual. However one must not forget that also the decoding, i.e., the reading and comprehension of such representations are among those mental abilities that are by no means natural to us and have only developed in the course of the development of communication.

And this is where the question about the role of history comes in. What are the individual components, the building blocks, out of which visual representations are assembled? This is a matter of searching for the origins of perceptual elements and for the codes and metacodes, the building blocks that we find today in user instructions. Codes are the ways in which the image and the real are related. In our pictures we draw lines and contours that objects in the real world do not have. We draw them in such a way that we can spontaneously recognize them again. Metacodes show us on a two-dimensional piece of paper elements that cannot be seen, such as motion and time; such building blocks make it possible to focus upon specific aspects and ignore others; they have arisen among various application contexts and target groups; some have become essential and others long forgotten.

2 Historic building blocks

Finding nothing in our images from the Stone Age, we turn to the tombs of the Ancient World. A fully developed pictorial culture arose here, one depicting human life on earth as well as in the hereafter, involving numerous artifacts. One generally finds no instructional texts here, the exception being conduct in the transcendental world. But whether such images simply depict activities or indeed give guidance or provide instruction is difficult to say. Images of technology and its use, however, do already exist in the Egyptian, Mesopotamian, and Assyrian cultures. Again and again one finds illustrations of humans, technology, and work.

Occasionally there are images that are clearly of an instructional nature. In 2000 B.C., in Mesopotamia, the floor plan of a house was scratched into clay; in the plan, the length and width of the rooms, drawn to scale, can be clearly identified. Plans are, as a rule, not always easily understood; the elements depicted tend to be strongly abstracted. Our example demonstrates that the relationship between the abstracted image and its source is more than 4,000 years old.

The site plan for a gold mine, a black line drawing with red script on Egyptian papyrus, was produced in 1300 B.C. (Figure 1). Three mountains are

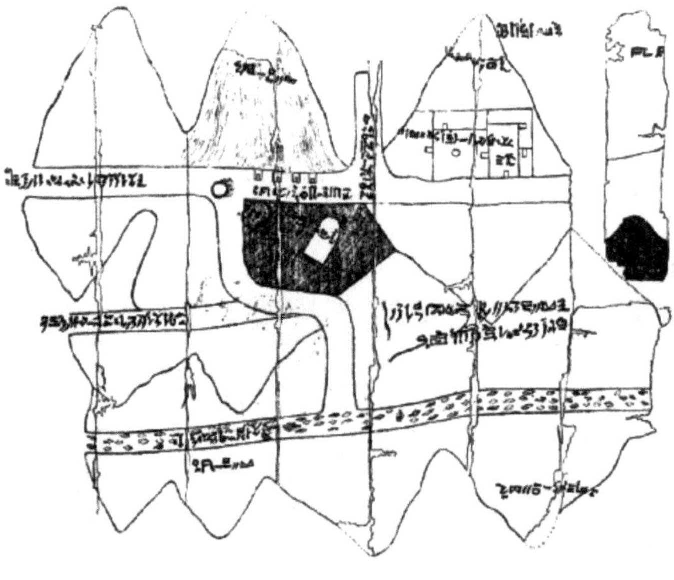

Figure 1: Site plan for a gold mine. Egyptian drawing on papyrus, 1300 B.C. (Egizio Museum, Turin, Italy)

visible above and four below. The map shows the path leading into the gold district in the area of Bechen, east of Koptos (Feldhaus 1959: 9). Mountains, valleys, rivers, roads, and settlements are drawn in; written comments are positioned so that they do not overlap the drawings. Images have thus been associated with verbal annotations from very early times. One thing was already clear to the designer of this plan: pictures alone are not sufficient; they need verbal descriptions to identify the individual elements. When no one is present to point a finger in explanation, this must occur using written language. This method of combining text and image, in common use today, is more than 3,000 years old. Picture detail and descriptive label are placed in physical proximity and thus establish a reference. A systematic method for designating components by means of a reference system has been known since the Middle Ages. When not enough room is available on a drawing for names, letters or numbers are added to details, which are then further explained in the text. The reference system initially found application in cityscapes (and therefore maps), and then also in technical illustrations.

At various times and in various cultures, different objects have appeared in our depictions of technology; by examining the images we can establish their relevance. Above all, it is in the presentation of war technology that one recognizes not only the trades and crafts of the time, but also the method of illustration. It is always interesting to observe the relationship, the fit, between humans, technology, and its use. The oldest known representation of a block and tackle can be found in an Assyrian relief from the 8th century B.C. (Lilley 1948: 23).

Its purpose was to supply the archers fighting on the battlements with munitions. The figures operating can be seen in the foreground (Figure 2).

This is in contrast to drawings from the 11th century, in which no figures are visible. But we do see the deployment of war technology. One can see the functioning of a flame thrower, for example, being used to heat a stone wall. The illustration is from the book Polioketika, by Apollodorus of Damascus, which records the development of war technology in the year 127 (Figure 3). Despite the lack of perspective in the illustration, the rupture in the masonry is clearly visible.

The illustrated manuscript Hortus Deliciarum by Herrad von Landsberg, the Abbess of Hohenbourg, at the convent of Ste Odile in Alsace, is not quite so militaristic, but it is an important milestone with respect to instructions (Figure 4). The 324 parchment pages with 636 miniatures were intended for the pupils in the convent; the work thus takes a pedagogic approach to presenting the technical knowledge of the time and therefore fulfills most of the criteria in our definition of an instructional text. The original pages in the manuscript were destroyed in a fire in 1870; only reproductions remain. Among them is an illustration of a wine press, which shows a water-driven mill (Feldhaus 1959: 14).

What makes the drawing interesting is that the draftsperson made no attempt to achieve a precise perspective, but rather tried to make visible those

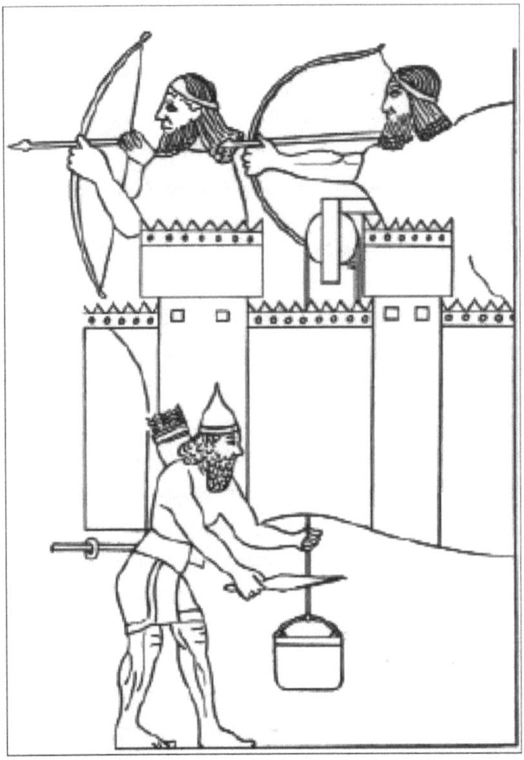

Figure 2: Block and tackle, Assyrian relief, 800 B.C (Lilley 1948)

important components that in a proper perspective would not be seen. A systematic approach to pictorial representation did not exist then; the draftsperson thus followed and made use of her own intuition, knowledge, and faculties.

Another example more clearly illustrates the problem: one of the most outstanding collections of technical drawings that exists for this period in the depiction of technology can be found in the travel sketchbook of Villard de Honnecourt, from 1245 (Feldhaus 1959: 14; Ferguson 1993: 81). It contains a drawing of a lumber saw bearing the following caption (in French): "In this manner one makes a saw, sawing through itself".

The allusion to a perpetuum mobile gives the device a greater technical significance, but was certainly not intended seriously. After all, the source of power can be clearly seen: a watercourse drives a crankshaft, which is connected with two wheels and two rods. The flow of the water (to view this properly the image must be turned on its side) is depicted in a lively manner. The water wheel is clearly powered from a natural source. Like the river, the

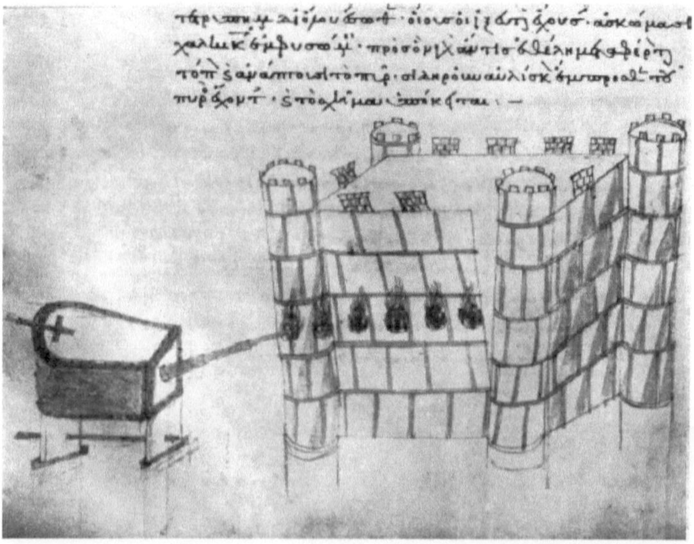

Figure 3: Flame thrower from the book Polioketika, by Apollodorus of Damascus, 1100 (Vatican Library in Rome, Italy)

return spring of the saw is depicted in a greater richness of detail, and the depiction of the fork reveals its woody composition. The vertical and horizontal attachments are shown in the perspective view that best lets them be seen. The entire setup, as depicted in the drawing, would not work; the saw, as well as the wheel's geared drive – intended to saw lengthwise through the log – are on the same axle; the feed rate would be much too fast. Likewise, if one looks at the position of the water mill's bucket wheel, one becomes aware of the crankshaft's direction of rotation; this would, however, rotate backwards. Apparently the draftsperson was not interested in these details. For him it was just a sketch depicting those elements that he found novel or interesting. This was not a matter of depicting an everyday machine, but rather of recording the unusual and curious. Important here was the operation or the replication, as suggested in the title: "In this manner one makes a saw." In technologically advanced cultures, educational travel and the resulting sketchbooks and travelogues served to scout out and then allow the reproduction at home of new technological developments (Figure 5). This form of early industrial espionage led finally to the development of patent agreements.

Many technical artists in this era did attempt three-dimensional views, but presentation of the significant components was always the first priority. Since the other parts are shown in proper perspective, the distorted representation cannot be due to a lack of skill on the part of the draftsperson. Ballstaedt

Figure 4: Wine press from the manuscript Hortus Deliciarum by Herrad von Landsberg, 1175–1195 (Rott and Wild 1944)

(1999: 77) refers to this as functional perspective and explains that the manner of representation has a didactic or methodological motivation: it shows all significant parts of the device in one drawing.

3 The view into the machine

Very few drawings of machinery from the Middle Ages have been preserved. Many documents were sacrificed to the church's destructive opposition to pictorial representation. The largest collection of extant works is from a self-taught artist known to us not only for his technical and anatomical sketches but as one of the world's most versatile artists. Leonardo da Vinci (1452–1519) left behind more than 10,000 drawings, which are carefully preserved in museums and archives and regarded as treasures of the greatest value. It is not only his draftsmanship that deserves our admiration; above all he solved many of the representational problems of technical documentation from the Gothic period.

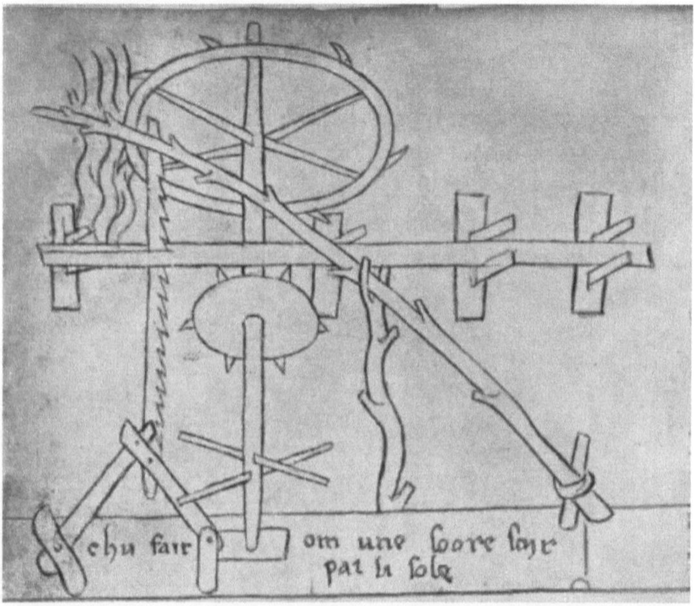

Figure 5: Lumber saw from the travel sketchbook of Villard de Honnecourt, 1245 (Bowie 1959)

Leonardo was not only a draftsman and painter, but also an inventor. His court appointment required that he provide instruction on technology to laymen. Many of his drawings give the impression that they were created while providing verbal instruction; his pedagogical experiences clearly influenced his unusual style. Indeed, Leonardo established the foundations of modern technical drawing: linear perspective, the use of proportion, and reference labels using letters and numbers were all carefully made use of and have been standard practice in drafting and printing since the Renaissance. Rather than force everything in his illustrations to fit into the formal perspective, he drew multiple sketches from various points of view and enlarged individual details. He also placed side-by-side several drawings, in the same perspective and size, of the same device such that the various states could be represented. The drawings could then depict before-after and if-then conditions, thus making it possible to illustrate a sequence of actions. While many of his contemporaries only depicted technology's outer appearance, Leonardo also depicted the workings of machines. This was a new way of seeing things. He viewed the machine not as being an indivisible whole to be captured in a single view, but rather as being made up of individual, elementary parts.

In 1480, Leonardo became involved with a representational problem that more than 200 years earlier had given the architect Willard de Honnecourt difficulties. Here one can clearly see the historical development. Leonardo depicts the entire structure in a single view; the machine is clearly drawn, to scale and in proportion. Individual details, however, are repeated once more for emphasis in a separate drawing. Thus the saw's automatic feed, with the ratchet and pawl that regulate the idle operation, are clearly recognizable. The functional perspective is therefore no longer necessary.

In 1500, Leonardo introduced the exploded view to the illustrator's arsenal of tools. He thus directs our view for the first time into the internal workings of the machine and technology. The exploded view, which displaces individual components from their actual positions and depicts them as if strung together

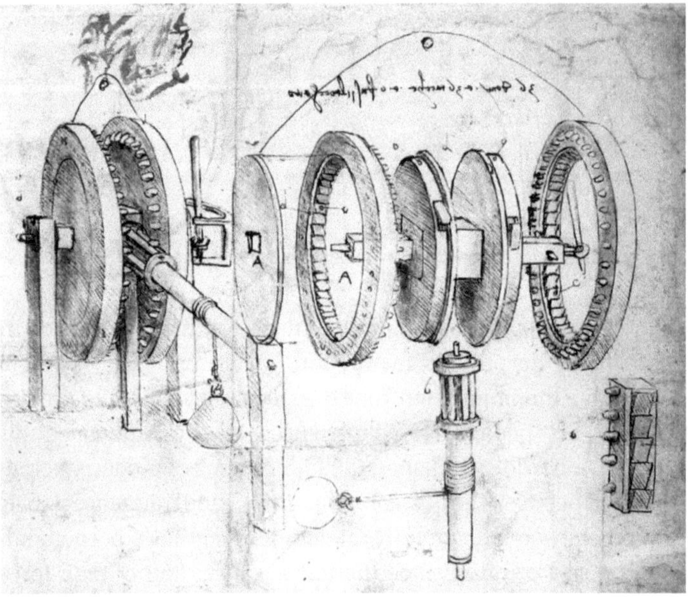

Figure 6: Exploded view of a gearing mechanism by Leonardo Da Vinci, 1500 (Codex Atlanticus, Biblioteca Ambrosiana, Milan)

on a thread, was perfected by Leonardo (Figure 6). It was now possible to recognize the manner and order in which the components were assembled. Today, even instructions for the simplest of hobby kits (Figure 7) make use of such diagrams, whose ancestry lies in the Renaissance.

Leonardo's technical drawings reflect the essential stages of development in modern technical documentation. These concepts are used today in the furniture assembly instructions from IKEA as well as the assembly manuals of aircraft maintenance engineers. There is one problem in characterizing Leon-

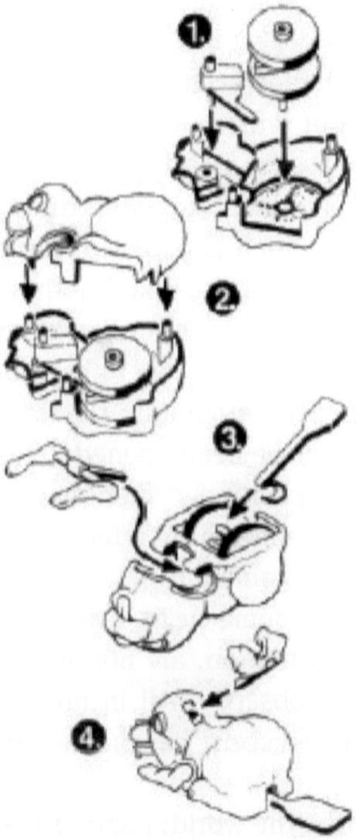

Figure 7: Exploded view in an instruction for a children's toy (Ferrero)

ardo as a pioneer in the preparation of instructional illustrations. An essential attribute is lacking: during his lifetime, none of his technical works was ever published. These are simply working sketches; however, it was his intent to develop public textbooks from his notes. His ingenious ideas must therefore be given due historical consideration. A further important representative of this period of technical illustration is Georgius Agricola (1494–1555).

Between 1533 and 1550 he published numerous books on mineralogy. His most important work appeared in 1556. In De re metallica he describes the principles and techniques behind the production of metals, from exploration to processing of the mineral ores. Well into the 18th century, the twelve volumes of his book served as the essential reference on mining and mineralogy. The contents included the use of wind and water as energy sources; the operation and function of ovens and smelting hearths, pumps and ventilators; the manufacture and refining of glass; the use of alum, sulfur and soda in the processing of mineral ores; also the administrative duties and obligations of mining officials and businesses and an examination of accidents and illnesses in mines. Georgius Agricola, doctor and mayor of Chemnitz, Saxony, described and depicted in various states not only the tools and work materials, but also the people who performed the work.

Unlike Leonardo, Agricola's drawings again depict scenarios in their entirety – not in a formal perspective, as earlier, but with a high degree of realism. These are not sketches, focused only on the essential, but rather carefully worked out artistic woodcuts (Figure 8). In contrast to hand-drawn sketches, they required more time and greater effort, which is reflected in the style. Landscape, tools, and workers have been carefully arranged and nothing is left to chance. Preliminary drafts were certainly available and followed in the production of the woodcut; thus the illustrations could be carefully composed. The woodcut gives us a view into the inside of the mountain; the tools are arranged for the benefit of the viewer. These, too, have been opened up as if by x-ray in order to reveal their functioning; they are positioned such that we can easily recognize the most important components. Labels make reference to the explanations of the terminology and thus to the commentary in the text. The figures depicted are busy with the equipment or, in some cases, even have time to carry on conversations. They are frequently in motion, are not always fully visible, and do not necessarily turn toward the observer. All in all, the pictures thus have a very lively quality, as if by chance a scene has momentarily caught our eye.

Human language is limited in its ability to describe the world; pictures can do this better. The preparation of the illustrations was so complex that the publication's release was significantly delayed. Agricola completed the manuscript in 1550, but he died in 1555 without having seen the completed edition of his masterpiece.

With Leonardo and Agricola we have the development of practically all representational techniques in use today. The use of color only first became feasible with the development of economical color printing methods; however this, too, had been applied under Agricola. In an early edition the flames of the fires and firing ovens as well as the cheeks of those standing nearby are emphasized in red. Today our visualization terminology is more precisely defined: it is possible to show components being illustrated in numerous directional views, as well as in phantom, cutaway, sectional and exploded views. Looking at the examples given, it is clear: there is nothing new about this – other than that standardization and new techniques have yielded images more modern in appearance. Axonometric projection, vanishing-point perspective, line-drawing conversion, and 2D- and 3D-visualization are some of the concepts behind these developments. The evolution of drawing instruments (particularly the introduction of the computer) and printing technology have greatly influenced the aesthetics of the pictures, even though the representational possibilities have more or less been fully explored.

Figure 8: View into the inside of the mountain from "De re metallica" by Georg Agricola, 1556 (Hoover and Hoover 1950, reprint)

4 New signs

New – and at the same time, not so new – in the development of illustrations is the so-called pictogram. They are new, in that they introduce visual signs into the text. They are at the same time not at all new, inasmuch as they were really our earliest language signs. Thus hieroglyphics belong to the pictograms; they iconically depict what they mean. Important is that unlike the phonetic alphabet, they do not make reference to a sound or articulation, but rather provide a specific sign that can represent an object, action, or function; they are the first language records. In that respect, there are possibly pictograms among the first cave paintings, and certainly one can point to the Chinese letterforms as pictograms. The relation between a pictogram and its meaning is sometimes very abstract; in some circumstances one is already speaking of symbols. And even pictograms must be learned; few are self-evident in their meaning. When the relationship becomes less clear, one speaks of ideograms. A picture of the sun can represent brightness, warmth, day, or daylight; this differs from the symbol, in which the commonality of image and meaning appears to have been fully lost. (Which sign represents the switching on of a device?) The development and cultural dissemination and acceptance of pictograms took a large step forward with the 1972 Summer Olympics in Munich. For the many arenas, sports facilities, public transit stops, and other venues it was necessary to find images that would not have to be translated into the many languages of the participating countries. This was certainly not a new idea. Otto Neurath, the Viennese master of "consistent visualization" – standardized methods of visualization – had already developed easily understandable images – he called them isotypes – for streets, train stations, airports, sports venues, hotels, department stores, and hospitals. In his 1936 book International Picture Language, he describes a visualization system designed to make complex situations easily understandable. "If one wants to make a quick and lasting impression, one uses images," is how Neurath summarized his philosophy. For the 1964 Olympic Games in Tokyo, the Japanese graphic artist Katsumi Masaru introduced a graphic signage system for the first time to identify the various sports. In 1967, Otl Aicher (1922–1991) and his colleagues were contracted by the organizing committee to come up with the visual appearance of the Summer Olympic Games in Munich. He further developed Katsumi Masaru's visual language, which was based on geometric figures intended to largely transcend cultural background. The resulting designs became a trademark of the Olympics (Rathgeb 2006).

The pictograms served as an internationally understandable information system on the grounds, in the Olympic Village, in the city traffic system, and at the other sports venues. Aicher's firm not only developed visual signage systems for large international events, he and his colleagues also worked for firms such as the electrical appliance manufacturer Braun, the lighting fixture manufacturer Erco, and the German broadcaster ZDF, in Mainz. Aicher was one of our most important and influential designers; and his influence is also to be found in the communication of technology in instructional materials.

There have been various attempts to standardize the use of pictograms; one example is the DIN 30 600 standard (Figure 9). An advantage of the picto-

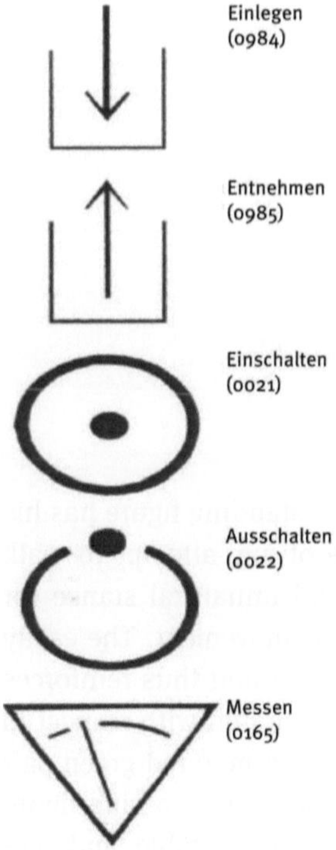

Figure 9: Pictograms from the DIN 30 600 standard (Deutsches Institut für Normung 1995)

gram is that it clearly establishes the relationship to an object or device. Particularly with those devices equipped with a display and complex modes of operation (such as a video recorder), pictograms can make it easier to find the correct setting. There is a direct correspondence between the display image and its reproduction on paper.

Following the German reunification, the Ampel-Männchen – the East German traffic signal figure representing the pedestrian – triggered a debate involving more than aesthetics.

Planned was standardization at the federal level of all traffic signage, with the practical result that the East German variations would be discontinued. This was an idea, however, that touched upon the emotions of the participants. Special interest groups formed to preserve the beloved figure. The West German version reduced the human figure to the essential elements necessary to make it understandable as a figure. The essential functions to be conveyed – walk and wait – are recognizable. Wait is represented by a standing male figure, in frontal view, his arms at his sides. The East German version goes beyond the

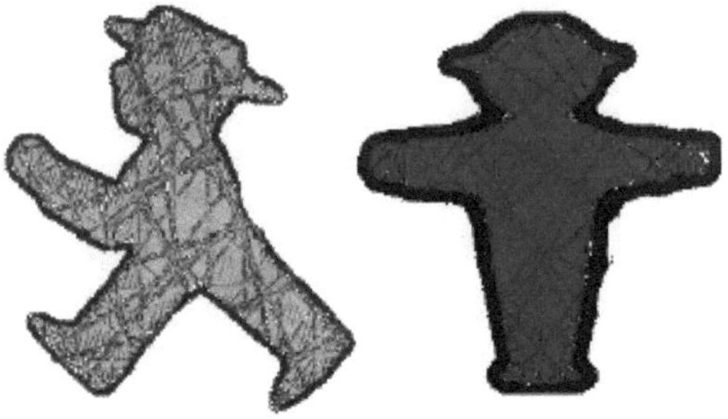

Figure 10: The East German Ampel-Männchen

abstract representation; it is livelier (Figure 10) and the standing figure has his arms stretched out wide – suggesting the hindrance of any attempt to walk and at the same time depicting an uncomfortable and unnatural stance for walking. The figure for walk depicts the dynamic of the movement. The easily recognizable profile of a hat suggests a hiker or wanderer and thus reinforces the figure's sense of movement. Both versions are quite similar with respect to comprehensibility, but the emotional appeal of the East German red-green pair is much greater than that of its West German counterpart. The walking figure is more energetic and lively; the standing figure is more stable and less ambiguous. Interestingly, we see here that pictograms can be quite lively and witty without ignoring the seriousness of the information being conveyed. In the end, they still tell us when it is safe to cross the street.

5 Innovation

Even though they were not publicly accessible in his own lifetime, Leonardo da Vinci's drawings are a part of the long history of the representation of technology and therefore of the history of instructional illustration. His quickly developed sketches were not intended for publication; they are preliminary stages in and notations for the development of a more elaborate composition.

Rough sketches have often served – not only in the case of Leonardo – as precursors for design drawings that were then published. These often then appear in unaltered form in the reading material of laypeople using technology – which thus makes them an aspect of this investigation. Sketches and textbooks as examples in the history of technical documentation have already been dealt with previously. One further important source is yet to be mentioned. In order for inventors to secure their legal rights, they must somehow protect their innovations; this happens in the patent office. England has had patent law since 1623 and the USA since 1790; the patent office in Berlin opened its doors in 1877. The registration for patent protection of a new invention requires that it be extensively documented. For technical devices, the documentation must include illustrations; these are then often copied with little or no alteration into the informational literature for laypeople.

With the sale of a new Mercedes Cardan, the Daimler Motor Company in Stuttgart-Untertürkhein also provided the purchaser with the vehicle's specifications and operating instructions. In the center of the instructional booklet from 1909 are several pages with drawings that only a professional would possibly understand (Figure 11). The annotations are as brief as can be – barely two pages of text. Under the heading "Braking" one reads: "The vehicle is equipped with two brakes, one of which is to be operated by means of the foot pedal C1 and the other with the hand brake lever D; the brake rotor of the footbrake is located on gear shaft F2, the handbrake acts upon the brake drum J on the rear wheels. To regulate the brake shoes, setting devices are located on the brake-shoe lever and the brake-shoe holder inside the brake drum of the footbrake and rear-wheel brakes, respectively" (Daimer Motoren Gesell-

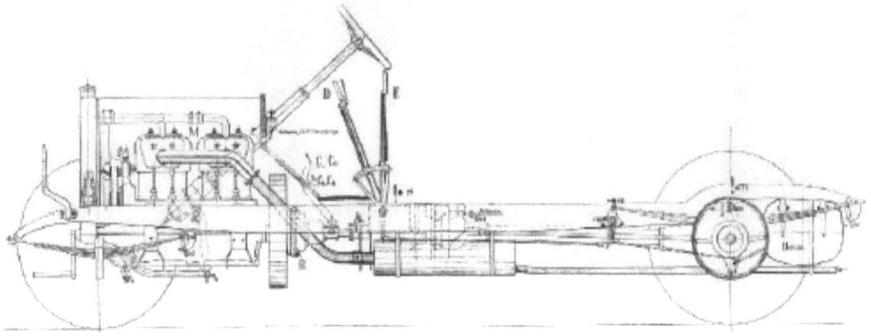

Figure 11: Specifications and operating instructions for the Mercedes-Cardan-Vehicle 1909 (Mercedes Benz Archive Stuttgart, Germany)

schaft 1909). Further paragraphs describe the steering control, speed change, reverse drive, pedal, handbrake lever, shut-off, and lubrication. One should not forget that the motorists of this time were technology freaks who were expected to attend to things that today would be taken care of by the authorized dealer or service facility.

A further example involves a technical invention that spread quickly, even among the slightly less-well-off middle class. We are speaking of the telephone, which can be traced back to the German inventor Philipp Reis; however, it was the American inventor Alexander Graham Bell who went on to develop and bring the technology to market. He submitted his patent application for an "Improvement in Telephony" to the patent office on February 14, 1876. As usual, the application included technical drawings that showed the interior workings of the transmitter and receiver. It was exactly these illustrations that appeared on the cover of Scientific American on October 6, 1877. Bell, however, neglected to patent his device in Germany. This was done by Siemens & Halske in order to secure the rights for the German Empire; they received the lucrative patent number 2355 for a "Telephone and Calling Apparatus with Magnetically Balanced Oscillator." Siemens' apparatus differed from Bell's in that Siemens replaced the bar magnet with a horse-shoe magnet. Regardless of the patents, one thing remains certain: the drawings that went to the corresponding patent offices went on to be published in professional articles and were even used in the user instruction materials (Figure 12). Portions of the design drawings

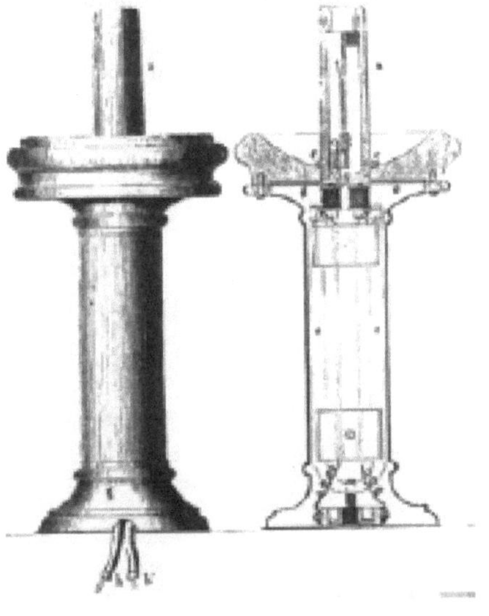

Figure 12: Telephones, Leipzig 1880 (Siemens Archive Munich, Germany)

appeared unmodified in the user instructions; the image on the right, in particular, is of little use to the calling party.

6 Motion and direction

Time becomes a major issue in the representation of action. Action, which, after all, is what instructions are really about, is certainly not static, but the pictures on a sheet of paper are. Various approaches to the problem of sequential action have been tried; three methods can be specifically identified:

1. Sequencing by means of changing image content
2. Motion lines within an image
3. Directional arrows within an image

When there are changes between two images of an object drawn from the same perspective, something must have occurred with respect to the object. The spatial proximity of the images and the unchanging viewpoint suggest a temporal change. Inasmuch as inanimate devices do not change of their own accord, a change of some kind must have occurred to them; by looking at and comparing the pictures, it is usually possible to discover this. At the same time, in order for a change to have occurred, there must be a cause; in the case of a set of instructions, this would be the operator or user. By comparing the images, the user can interpret the sequence and bring a device from its current state to a new one, corresponding to or at least more closely resembling the subsequent image. Such image sequences can be found – what a surprise – in the work of Leonardo da Vinci; one well-known sequence of drawings depicts his flying machine in various stages of operation.

This method was perfected in another words-and-pictures medium – one which might not occur to the technical editor hard at work, namely in the comics. It is worth taking a look at the techniques developed in this medium, which goes back more than a century. The year 1897 – the year in which Rudolph Dirks' series the Katzenjammer Kids first appeared – is considered by many media historians to be the beginning of the comic strip. The series bears a striking resemblance to the illustrated stories of the two troublemakers Max und Moritz, published by Wilhelm Busch (1832–1908). In each, two lines of text elucidate the pictures, yet the narrative elements are also understandable without the text. The layout and sequencing of the images is important in this respect. We have a sequence of pictures whose perspective and framing remains unaltered, but not so the individual objects within the frames. When,

Figure 13: Busch, Wilhelm: Max and Moritz, 1965: 25

for example, in the fifth Max und Moritz trick the two rascals place May bugs beneath Onkel Fritz's bedspread before he retires to bed for the night, only the uncle, the beetles, and the blankets are in motion (Figure 13).

Nothing else changes. From these individual pictures and their largely minimal variations we are able to construct a sequence of time and action. What is exceptional about the comics is not the interpretation of the images, but rather the creative interpolation of time between the recorded moments.

One of the oldest methods of representing motion within a picture is the use of lines following the direction of movement of objects in motion. The specific objects that are dynamically involved, their point of origin, and their direction can thus be visualized. Such motion lines can be seen on a shard dating from the 13th century B.C. that was found in Deir el-Medineh, where the workers lived who were building the royal graves on the Thebes (McDowell 1997: 80). Lines have been painted in front of the mouth of a bending figure;

Figure 14: Shard Deir el-Medineh, 1300 B.C. (McDowell 1997: 80)

alongside are the words "blowing into the oven" (Figure 14). Scarcely any artist using motion lines would be familiar with this fragment from Ancient Egypt, particularly as it was excavated only a few years ago. This same metacode, however, can be found almost without alteration in an illustration from Superman (Figure 15). Such "blow lines" originating from the mouth indicate time and direction. These examples show us that the method has been frequently reinvented in the course of history and that it is an entirely natural way in which to represent motion. Of interest is also a picture by a seven-year-old boy. Asked what he had painted, he replied: "A monster raising his leg." And once again the lines are in the direction of the motion. What is exceptional about the picture is that the boy very likely had little or no experience with comics. The picture was created before 1989, in Dresden. Perhaps the local telephone book provided a model: its 1967 edition displays an excavator with a dynamically moving shovel. Today it is not possible to determine whether the young boy from Dresden had access to Western comics, or the telephone book, or whether he discovered this metacode himself. What is

Figure 15: Andy Warhol's Superman (1961), from the collection of Gunter Sachs

certain is that it is easily understood, and there is further evidence for this thesis.

As a predecessor of today's comics, Wilhelm Busch, too, drew line strokes to represent motion. In Max and Moritz's third trick, they set upon a footbridge with a saw: "Ritseratse! Such deceit / beneath their feet, the bridge grew weak" (Busch 1965: 18). After Meister Böck, the tailor, has fallen in, two geese pull him out. Motion lines here represent not only dripping water – something for which such lines are not especially suitable – but above all the dynamic movement. This becomes even clearer in their next trick, in which the spiteful boys place black powder in the tobacco pipe of Master Lämpel. As this explodes ("Boom! – The pipe goes off") Busch again draws lines, radiating in all directions. The moment of the explosion is temporally captured and at the same time, by means of the motion lines, drawn out. One sees the source of the movement and simultaneously the results, as teacher, pipe, and all fly into the air – "all go up in a flash of light" (Figure 16). This method also found its use in a sign warning of danger (Figure 17) – most likely without any knowledge of its earlier creator.

Figure 16: Busch, Wilhelm: Max and Moritz, 1965: 22

Figure 17: Classification, packaging and labeling of dangerous substances, Council Directive 2001/59/EC for explosive substances

With the help of motion lines, one can fully represent complex movements, inasmuch as the lines do not necessarily need to run straight. The history of comics offers numerous examples of this method. One can probably open most any comic book and see other possibilities and variations in style. This form of representation has only one restriction: it depicts objects where they are at the moment recorded and where they were previously. The lines trace the direction of an object's movement, but do not show us its previous state. It is also not possible to identify continued movement or the future resting point. These limitations are resolved by means of another pictorial convention.

In his dissertation "Presentation media for product interaction" Piet Henderikus Westendorp (2002) looks at the history of the directional arrow and made some interesting discoveries. The first use of the arrow for symbolic purposes was by the Ancient Greeks, who used it in their compass charts to indicate where north should be. What was very likely the first use of directional arrows to indicate the direction of flow of a body of water can be seen in maps from the end of the 18th century. Westendorp dates use of the first directional arrow in a technical illustration to 1737, where it is used to show the flow of a brook in order to make clearer the direction of rotation of a water wheel (Figure 18). At this time it still had a barbed tip and feathers at the end of the shaft. The origin of the symbol is clearly recognizable. The arrow is aimed at a target. It thus designates direction and movement, the exact location of the target, and the time, specifically the future. Its representation has in the course of time been further schematized. At the beginning of the 20th century, we find the arrow replacing the pointing finger that was previously in common

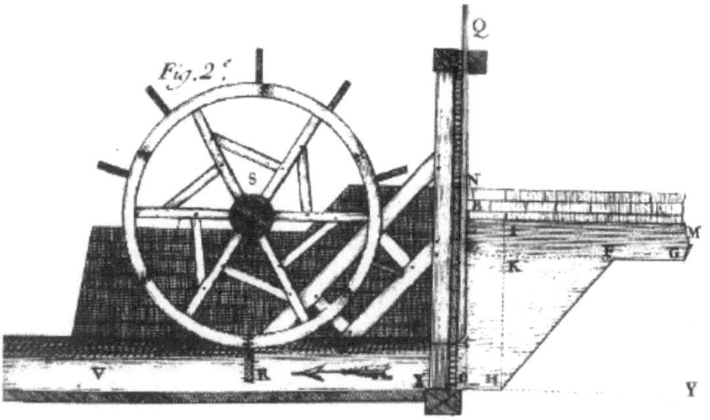

Figure 18: Water wheel with arrow that indicates the direction of rotation in "Architecture hydrolic" by Bernard Forest de Bèlodor, 1737 (Westendorp 2002: 57)

use in traffic signage. In modern instructional illustrations, the arrow has still further meanings. Ballstaedt (1999: 86) augments their use with functions "for the representation of cause-and-effect, temporal direction, and material and information flows as such as vector, pointer, and token of movement." Above all, the arrow is used as a code for such elementary actions as pulling, pressing, pushing, and turning; it is also worth noting that in the course of history the arrow has replaced the human figure as the active agent in visual instructions.

For the presentation of motion in technical illustrations, the arrow has more or less taken over. It is more common than motion lines, as the arrow can show us the current state as well as the direction in which a movement is to take place. The arrow is one sign, but consists of two components: the shaft and the tip. While the shaft, as a line, shows us the progression, the tip, as directional indicator, points to one of the two sides (Luhman 2002: 72); one draws a motion line and then with the addition of the tip creates a temporal differentiation from the current state; thus by means of the arrowhead the direction of motion in the future as well as potential or optional positions can be indicated.

7 Pointing and indicating

The simplest way to point to something is with the finger. Even small children just beginning to master the spoken language use the extended arm and finger to point: at things they want or whose name they want to learn. At this development stage, there is a close relationship between language acquisition and pointing.

Pictures, too, also draw attention. Symbolic uses of the forefinger have also developed, and the arrow also plays a role in this context. We follow its direction and our view is drawn to the area where the arrowhead leads. In general, it can be said that we easily give our attention to things that can be held in the hands. A further technique has been borrowed from the comics and cinema. Once the context of an object of significance has been established in a drawing, it can be depicted in an enlarged view in the next drawing. The effect is as if the camera or observer has moved in for a closer view. Metacodes can also be useful in making visible the relationships between object and context. Circles indicate an area of interest and larger circles enlarge the scope. Details can then be seen as if viewed through a magnifying glass. Emphasis can also be provided by metacodes known to us from the comics as speech balloons. The tail points to the object of significance and the balloon displays relevant and enlarged details. With the speech balloon in the comic book just as with the technical illustration, it is a matter of origin and content.

8 Looking ahead

Leonardo da Vinci, Georgius Agricola, Wilhelm Busch, and Otl Aicher have all contributed to the historical development of technical illustration, each in his own unmistakable way. Each has played an important role in the development of visual instructions. It might seem that such development has now reached a conclusion. Many of the principles of drawing have long since been known; a brief look into the development of some of the historical building blocks has shown us this. What is still to come? Illustrations depict an object in specific states. The graphic conventions we use, the codes and metacodes, are a part of a long, ongoing cultural history. They are thus subject to cultural developments and will continue to develop new forms themselves. Whether in drawings on papyrus or modern glossy printing, various tendencies can be observed: first, the codes are based on and change according to the artistic conventions of perspective, detail drawing, and the human figure; second, the development of new drawing instruments and mediums significantly influence that which can be illustrated. Color blends first became reproducible with the introduction of computers and the appropriate software; the use of color is also a question of the printing process and its costs. Changes in this area have been and continue to be rapid. The metacodes – among them the representation of space, time, and motion – are associated with learned and reproducible methods, which draw upon various cultural sources. Graphic characters or symbols are a part of many cultures, but their specific shapes and forms are subject to conventions and they are therefore only understandable within such contexts; here, too, change will certainly continue. Above all, the simulation of motion and three-dimensional space is technically far from being fully developed. In an age that is just now evolving from a paper-based to a video display-based culture, new methods of representation are certainly to be anticipated. Modern GIF animations and Flash presentations are early heralds that can only hint at what is still possible and what will soon become routine or even old hat.

References

Aebli, Hans
1981 *Denken: das Ordnen des Tuns, Band II: Denkprozesse*. Stuttgart: Klett-Cotta.
Ballstaedt, Steffen-Peter
1999 Die bildliche Darstellung von Handlungen, in technischen Dokumenten. In: Schwender, Clemens (ed.), *Zur Geschichte der Gebrauchsanleitung*, 67–94. Frankfurt/Main: Peter Lang.
Bowie, Theodore
1959 *The sketchbook of Villard de Honnecourt*. Bloomington: Indiana University.
Busch, Wilhelm
1965 Max und Moritz. In: Busch, Wilhelm: *Humoristischer Hausschatz*, 11–32. Köln: Buch und Zeit Verlagsgesellschaft.
Daimler-Motoren-Gesellschaft
1909 *Beschreibung und Anleitung des Mercédès-Cardan-Wagens*, Nr. 10/40. Stuttgart-Untertürkheim (Mercedes Benz Archive Stuttgart, Germany).
Deutsches Institut für Normung
1995 *Graphische Symbole nach DIN 30 600. Bildzeichen Übersicht*. Berlin: Beuth Verlag.
Feldhaus, Franz Maria
1959 *Geschichte des Technischen Zeichnens*. Wilhelmshaven: Franz Kuhlmann.
Ferguson, Eugene S.
1993 *Das Innere Auge. Von der Kunst des Ingenieurs*. Basel, Boston, Berlin: Birkhäuser.
Hassenstein, Wilhelm
1941 *Das Feuerwerkbuch von 1420*. München: Verlag der deutschen Technik.
Hoover, Herbert Clark and Hoover, Lou Henry
[1556] 1950 *De Re Metallica. Georgius Agricola, translation from the 1st Latin edition of 1556*. New York: Dover.
Lilley, Sam
1948 *Men, Machines and History*. London: Cobbett Press.
Luhmann, Niklas
2002 *Einführung in die Systemtheorie*. Heidelberg.

McDowell, Andrea G.
1997 Die Schriftkultur einer altägyptischen Siedlung. *Spektrum der Wissenschaft* 1997/2: 76–81.
Ratgeb, Markus
2006 *Otl Aicher* (English edition). Berlin: Phaidon.
Rott, Hans-Georg and Wild, Georg (eds.)
1944 *Hortus deliciarum. Der „Wonnen-Garten" der Herrad von Landsberg. Eine elsässische Bilderhandschrift aus dem 12. Jahrhundert.* Mühlhausen/Elsass: Braun.
Schwender, Clemens
1999 Die Gebrauchsanleitung – eine Anleitung zum Gebrauch. In: Schwender, Clemens (ed.), *Zur Geschichte der Gebrauchsanleitung*, 1–26. Frankfurt/Main: Peter Lang.
Westendorp, Piet Henderikus
2002 *Presentation media for product interaction*, Delft: Technische Universiteit Delft.

III. Relationships between technology, communication and culture

Vasco Alexander Schmidt

Software documentation: The role of computers as a means, tool and medium

1 Introduction

Reading, writing and arithmetics: these abilities are usually referred to as cultural techniques. Only he who is able to read, write and sign is involved in the exchange and development of new knowledge and can only then really be a member of our industrial and knowledge-based society. However, according to cultural reading, writing and being numerate is no longer sufficient nowadays. New cultural techniques, such as the handling of computers (Krämer and Bredekamp 2003: 16), emerged a long time ago. Computers are utilised to control virtually all economic and social processes. They represent a substantial part of our daily life. In addition, mainly within the sector of mathematical sciences, they act as cognitive instruments. Due to the fact that the number crunchers, as computers were referred to in the beginning, have been used for complex simulations and experiments for a long time, they also convey new scientific insights (Borwein, Bailey and Girgensohn 2004).

The theory that the use of computers represents a cultural technique emphasises the great significance computers have in today's world and offers numerous connecting factors for evaluations of philosophical, sociological and linguistic nature. Seen from the perspective of technical editors, the thesis appears exaggerated. How can something that usually requires extensive manuals represent a cultural technique? Books do not require manuals. Reading, writing and arithmetics is merely learnt once and is then mastered for all times. New software is consistently – and always in a different manner – difficult to operate. It can actually not be managed without tutorials, help functions, FAQs, function lists or fixed procedures. In spite of this, many IT professionals think that computer systems will become self-explanatory in the future – and not longer in need of documentation. A belief that is certainly considered to be a myth by software documentation experts (Mehlenbacher 2003: 528).

This polarity, emerging through the view of cultural studies and clashing with the practical experience standpoint, leads to the core of the documentation problem: How can computer systems be developed that can be operated intuitively? How the documentation expenditure be reduced via a good design of the computer systems? How could one write computer documentations actually useful for the daily work and worth reading for the operator?

Studies have revealed that, up to now, documentations are hardly read and do not really meet the demands of the readers. So far, a clear picture of how documentations are actually used and how they can be improved has not yet emerged.

This paper intends to elaborate the core of the documentation problem from a semiotic viewpoint. This perspective is based on the thesis of the culture technique of computer handling. First it is intended to trace back how the profession of technical editing developed and which topics in this profession have gradually gained significance in the course of time. Subsequently follows a description of the challenge of software documentation from a linguistic view. In short, this is represented by the fact that there are no conventionalised text types technical editors could orient themselves on. For this reason new strategies for the description of software are repeatedly necessary. Following the discussion of the subject matter of software seen from a semiotic view, the existing manual and text book literature is introduced on an exemplary basis and the strategies of the software documentation specified there are newly classified from a semiotic perspective.

2 Software documentation

Software documentation can be perceived as the appropriate and recipient oriented creation of information offers via software products. Depending on the underlying definition of software, software documentation reaches from the comments in the coding to guidelines for the installation and maintenance, operator manuals and E-learning units right to the terms at the user interface. The main emphasis is directed to texts printed or displayed on the monitor, even though this also includes non-verbal information carriers such as images, animations and videos.

In the definition underlying here, software documentation represents a sector of technical writing. Due to its subject matter, however, it disposes of an own profile. In regard to its complexity, software may be compared with control systems of major industrial plants for which extensive documentations are also compiled. Software products are, however, more abstract as other technical artefacts and are initially mediated via media, and therefore only visible as artefacts in the course of handling and through description. Subsequently, software connects a technical domain with the application domain in a particularly complicated manner.

2.1 History

The documentation of computer systems is as old as the computer itself. However, the demand for software documentation did not increase as late as the 1980s as personal computers and work station computers were developed and programmes for word processing and spread sheets came into existence. In the same period of time, the profession of technical editors, who professionally produce user manuals for technical equipment, also emerged. A significant influence on the professional image, according to general opinion, was excerted by the then also fast growing sector of software documentation (Göpferich 1998b: 3).

Since the 1980s, extensive guidebook, textbook and research literature on software documentation has been generated. Even though the objectives, basic conditions and methods of software documentation are herein defined and theoretically founded in a very different way, similarities and development directions are noticeable.

Whereas the emphasis was initially directed to the documentation in printed form, the availability of hypertext and internet technology starting in the 1990s facilitated the emergence of online documentations so that separate guidebooks are dedicated to this topic (e.g., Brockmann 1990; Horton 1994). In particular the documentation of web-based user interfaces is emphasised in guidebook literature (e.g., Kilian 2007). As a consequence, the research on usability, i.e. the user oriented design of user interfaces, and the research on Man-Machine-Interaction in general, has become more active. New courses of studies for technical editing and communication are developing, which is also further enhancing the professionalization within the sector of software documentation. Finally, textbook authors intentionally attempt, in addition to practical experience, to also increasingly involve academic findings in their presentation of technical writing (e.g., Göpferich 1998b) and software documentation (e.g., Barker 2003). Guidebooks without any reference to academically founded theories are, however, still being published (e.g., Bremer 1999).

2.2 Research issues

The main focus of attention of the research and development of software documentation is represented by the desire to improve the documentation. The purpose of the research is to examine how documentation can be created, managed, and displayed most efficiently, whilst ensuring its relevance to users.

Great attention is directed to the tools for technical editors in order to design the creation and presentation of documentation more efficiently. In the

1990s, apart from the embedding of documentation into the software and context-sensitive communication aids, the usage of hypertext and the separation of texts and their appearance (e.g., via XML) was the centre of attention. From the year 2000 on, the concept of Single-Source-Publishing has increasingly been addressed. This concept attempts to uniformly file source contents to subsequently reassemble them flexibly for various documentations. The ideas of the Open-Source-Software are also assigned to the software documentation to directly involve the users in the creation of the documentation.

Several research projects are dedicated to the technical conditions of writing. Within the scope of the Single-Source-Concept, strategies out of the field of rhetorics and findings out of the sector of linguistics in terms of cohesiveness and coherency are addressed to divide documentations into smaller text components easier to reuse. Empirical studies involving methods out of semiotics and text linguistics try to draw conclusions directed to the optimal structure of documentations and have lead to a multitude of individual results, e.g., for the design of indices, step-by-step manuals and to the usage of screenshots.

In the course of testing the quality of software documentation theories out of psychology, didactics and dialogue research are applied. Ethnographical field studies – so-called workplace studies – are utilised to attempt to register the everyday use of software. The objective of modelling such application scenarios is to support the technical editors in designing a better documentation for the respective target group. Other research issues address the organisation of software documentation creation, e.g., the cooperation among technical editors themselves or with users, support staff, and software developers.

2.3 The main challenge

Software documentation is in fact doubly influenced by technical progress. On the one hand, the operation mode of software itself is consistently changing. Due to the fact that the object in question and its application are changing, new strategies for their description are in constant demand. Therefore, the software documentationis forced to consistently adjust to new technological conditions. On the other hand, documentation tools also develop themselves further, which leads to the continual development of new word processing formats, forms of storage, and presentation of the documentation. As a consequence, this in turn continually requires new concepts for the creation and design of software documentation.

It is therefore no surprise that the history of software documentation, from a technical point of view, does not show any constant characteristics. This similarly applies to the text of software documentation. From a linguistic perspective – contrary to for example journalism – they hardly include conventionalized text types. Merely, in rare cases, can technical editors fall back upon application standards, in particular there are no proven text samples (Schmidt 2007). Consequently, from the linguistic perspective, the main challenge of software documentation therefore comprises the development of strategies for the description of software in terms of a prescriptive linguistic and to operationalize these for the creation of texts.

3 Subject matter software

The term Software – even though subject matter of the software documentation – is mostly not clearly defined in guidebooks, text books and research literature. Implicitly software, by international and national standards, is perceived as "intellectual property", respectively an "immaterial product", "which consists of programmes, procedures, rules and all respective descriptions of an information processing system". In addition, software is "independent from the media [...] it is stored on" and "[...] only functional in cooperation with (operation) software and hardware" (Deutsche Industrie-Norm DIN 66230 from 1981).

From this perspective, software is differentiated from hardware and the operating software. This is done to limit the subject of the software documentation to the actual software intended to be described. Indeed, in practical experience, the differentiation of hardware, software and operating software is certainly not always clear. There is also software closely associated with hardware, which is, for example, intended for controllers of drives or input units. These software components can only be described in connection with the physical equipment components. Operating systems and programming languages require an own documentation. Last but not least, within any software documentation – at least implicitly – hardware is of significance, as software, in the end, is operated via the Man-Machine-Interface of the computer system on which the software is active.

This is possibly the reason why the differentiation between hardware and software cannot be found in the publications of semiotics, philosophy and media studies. There, the terms computer and computer system are used. Unlike regulations for software documentation, they merely mention computers and computer systems. The terms software and hardware which are present in standards for software documentation do not appear.

The use of the term *computer system*, which comprises the hardware and software without further differentiation, is an attempt to handle the nature of the subject matter in a uniform, global scope. Proponents of semiotics, such as Nadin (1998) therefore strive for the development of a meta-program "that unifies software, data flow, input/output performance, connectivity, process and human interface, cultural and social acceptance, and learning" (Nadin 1998: 137). This challenge could be mastered through the "integration of computers into human pragmatics in order to make them appear as extensions of human intellect and skill" (Nadin 1998: 136–137).

This motive of embedding computer systems into the environment of man and society is evident in all approaches used in the fields of arts and social sciences regarding computers. However, this aspect is certainly less evident in the publications on software documentation. In these publications, the emphasis is on the description of a technical product, even though awareness of the necessity of taking a broader approach has increased in recent years.

3.1 Semiotic machines

When examining computer systems from the semiotic perspective, the following sectors are differentiated (e.g., by Nadin 1998: 136): on the one hand the procedures within the computer system, i.e. all types of processing (computation), and, on the other hand, the interaction of individual persons with computer systems (interaction), as well as the interaction of people with each other by means of computer systems (networking). The differentiation of these three sectors already indicates that computer systems are special machines. Other machines – be it a loom, vehicle or a bottling plant for beverages – are seen in a completely different way. They would not be associated with internal computation processes nor the facilitation of networking amongst people.

From a semiotic viewpoint every technical artefact which is deployed for a specific purpose can in fact be conceived to be semiotic, as – using the terms of semioticians – it represents its use through its realisation (it is a signifier that signifies its use). A hammer signifies through its appearance that it is suited to place a nail in the wall. In the case of cars, it is easy to perceive that they are suited to transport people. Computer systems are also technical artefacts that were developed for a specific purpose. One important characteristic is, however, that it is not possible to judge the performance of computer systems on the basis of their form or appearance.

The performance is all the more the result of the programming with the support of symbolic languages. Computer systems do not operate with physical but with purely abstract objects. Their strong point is the reproduction of a limited sector of reality on to a system of signs. In a certain sense, the depicted signs have subsequently separated themselves out of their original context and display an abstract image of this application context. This however enables their use in other algorithms and their transferral on to other application contexts. In this sense, computer systems, in regard to their internal structure, are equal to mathematical formalisms (Krämer 1988). This observation supports the theory that computer systems represent semiotic machines (Andersen 1997; Andersen, Hasle and Brandt 1997).

This abstractness is accompanied by two essential consequences. Computer systems are able to adjust to our social world in a special manner due to their abstractness, as this is characterised by a multitude of signs and symbols, sign processes and interpretations. A great part of the working environment and private life is already dominated by computer systems and this influence is constantly spreading over further sectors (e.g., since the end of the 1990s: internet book stores, the electronic supply chain management or online dating).

Simultaneously, the abstractness of computer systems leads to a certain invisibility. Computer systems are integrated into technical devices such as clocks, telephones, CDplayers or cars. Here they are embedded systems which, at the most, reveal the ingoing and outgoing data. The internal procedures remain concealed.

Even a look into the interior of computer systems does not necessarily help to understand the mode of operation. One reason is that they cannot only be seen as objects initiating sign processes when users interact.

According to the perception of semiotics, computer systems themselves are active participants of such sign processes. A typical example is the compiler, which generates an executable program out of a program code, therefore interpreting the used programming language. In their interior, computer systems dispose of a multitude of sign systems and sign processes and levels of abstraction which can solely be utilised through a technically initialised interpretation. From a general point of view, in everyday life, our expectation also assumes the computer system to be the active participant in sign processes. This happens, for example, when operating a program. Here we refer to the "computer" to "ask" or "it" has "made" an error, or "it" has "misinterpreted" something (Andersen, Hasle and Brandt 1997: 552).

The semiotic viewpoint to computer systems already displays its particularity at this point: they are semiotic machines characterised by the use of abstract sign systems therefore being universally applicable, which in turn leads to their application within almost all sectors in life. Computer systems are deployed in a whole range of various scenarios – whether it be as an invisible black box in a technical device or as an active participant which is treated as a person.

3.2 Tool and media

As computer systems are machines deployed for a specified purpose, they can also be referred to as tools. They are, however, tools of a particular type. On the one hand, they are able to support operating processes fully automatically, like the accounting of major companies or the account management in banking. On the other hand, they also provide us with support in the course of intellectual activities such as calculating and writing.

The rank of a cultural technique is, however, assigned to computer systems for another reason: they cannot only be perceived as a tool, but also as a medium. In a narrow sense, media can also be interpreted as a means to support communication. The internet technology, for example, gives us the opportunity to cooperate in teams over long distances. The physical absence of the participants in the course of such a computer-mediated communication does not only support the conventional cooperation of people, but also leads to a new form of human conduct. This conduct of the people displays computer systems as a medium in a vastly more extensive meaning. In internet forums, for example in the course of chats, it is possible to adapt new roles, to adapt new names and even new, virtual identities. The participants then act symbolically.

Within all these forms of communication, the involved parties – even though with gradual differences – exchange communication amongst each other in such a way that they often do not perceive the influence of the computer systems on this exchange. The computer is virtually transparent to them. This transparency, subsequently the non-perception, is considered to be an essential aspect of media (Krämer 1998: 75). Traditional tools do not, or only in a very limited scope, dispose of this characteristic.

Both perceptions – the computer as a tool and the computer as a medium – are very different in their basic nature but do both always simultaneously claim validity and are imminent at every contact with software. For this analysis, the characteristic of the computer systems as media is decisive since, in the first place, computer systems can only act as cognitive instruments due to their characteristic as media. In science, computer systems are specifically deployed for this purpose to see matters which would hardly or not be conceivable in any other manner. Computers represent an aid to discover the world.

3.3 The necessity for explanation

Anyone taking a closer look into the coding of a computer system encounters an unmanageable interaction of signs disposing of numerous cross references. The coding of computer systems presents itself to software developers, technical editors and users from respectively individual perspectives and is accordingly interpreted differently by them (Andersen, Hasle and Brandt 1997).

The user interfaces, on the other hand, offer a less abstract image of the software. As initially mentioned, one of the myths of software documentation is the perception that software documentation will become obsolete in the future (Mehlenbacher 2003: 528). This is founded by the argument that user interfaces will be totally transparent in the future and the software subsequently easier to handle, if not even operable on a self-explanatory basis. As a consequence, any additional instruction regarding the handling of software would become unnecessary. Such a perception is supported by the experience that the handling of software is learnt by using it and that this would, as in the case of a media, lead to its invisibility.

From a legal point of view, software documentation will surely always be necessary as, according to the currently valid regulations, a technical device may only be marketed in combination with a documentation comprising the functional range as well as an instruction on the appropriate handling.

However, regardless of such legal issues, computer systems will always remain to be in need of explanation due to their characteristic as semiotic machines. All types of sign systems leave behind traces which can be the cause of new interpretations and consequently understandability processes. Media academics refer to non-intended distinctive features, so-called pre-semantic phenomena (Krämer 1998: 79). In the course of the handling of computer systems such semiotic traces also appear and can lead to false perceptions and operating errors.

In addition, the areas of applications are by far not yet wholly utilised. Computer systems are consistently infiltrating new areas of the economic and social life. Due to this fact also the handlings of computer systems keep changing. Consequently, the manner we perform our tasks with computer systems, be it on a private or on a professional basis, is also consistently changing. The application of computer systems in turn affects the respective areas of application and partially restructures them.

Computer systems, as semiotic machines – in spite of their character as media and their ostensible transparency regarding their handling – will always remain in need of explanation.

3.4 Computer systems within the semiotic triangle

In order to make the characteristics of software more perceivable, the sign model of Ogden and Richards [1923] 1989, which is often used in linguistics, is quite suitable. It identifies three dimensions arranged in a triangular order. The linguistic sign is located on the left hand side; on the right is the extra-linguistic reality the linguistic signs are referring to. Located above is the cognitive perception we associate with the linguistic signs, respectively the extra-linguistic reality.

If one examines computer systems with the aid of the semiotic triangle, the computer system is located at the left corner point (see figure 1). As a semiotic machine it is, in a certain sense, equal to the mathematical symbolic language, thus respresenting linguistic signs in the triangle. Located at the right corner point is the application domain presented by the computer system. This corner point represents everything modelled via the sign language in the interior of the computer system. The third dimension is represented by the cognitive perception of the operator using the computer system. Even though there is a direct connection between cognitive perception and reality, the user does not only access the reality directly but also via the cognitive perception of the application domain, which he was mediated by the computer system. The same applies vice versa: The user also forms a cognitive perception of computer systems based on his cognitive perception of the application domain.

In case of business software, this phenomenon is clearly visible. A novice in accounting will also utilise an accounting software to increase his knowledge in respect to the processes and regulations within accounting. Experienced accountants dispose of such profound knowledge within their area of specialisation so that they will subsequently rather be able to recognise the limitations of the respective software in the presentation of an ideal accounting process.

Now the identified characteristics of software can be illustrated with this model as follows: if the user applies the computer system as a tool, he performs tasks on the application level and therefore intervenes into reality via his cognitive perception with the support of the computer systems. If he, mediated by computer systems, forms a cognitive perception of reality, the computer systems, in contrast, act as a medium.

Both of these avenues to reality, as generally known, are conducted via the user interface of the computer systems. Thus, the users control software and utilise them as a tool. At the same time, the user interface is the cause of the sign processes, to which the users form the cognitive perception. If users apply the computers on an intuitive basis, this handling is equal to talking in a natural language. Usually, however, the handling is not intuitive because – as clarified above – computer systems are in need of explanation. The user interfaces, in this case, are then repeatedly the cause of irritations due to their design. The user is then forced, either deliberately or not, to form a cognitive perception of the computer system, the reality as well as the connections between them. The causes for the irritations are the pre-semantic traces mentioned above in Section 3.3.

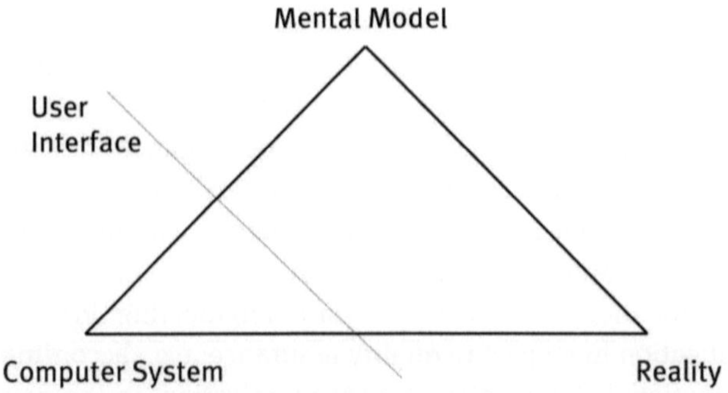

Figure 1: Computer systems in the semiotic triangle

4 Strategies of software documentation

The semiotic program mentioned in the introduction requires a standardised analysis of computer systems and their application. The objective of such an analysis was equally expressed pragmatically (e.g., Nadin 1998): It is supposed to make aware of sign processes and – above all – help initiate sign processes in a manner enabling people to master the intended tasks with the help of computer systems. In textbooks and manuals of software documentation such a program is nonexistent. What can be found, rather represents individual strategies that do not always fit together to an overall picture but offer links to the semiotic approach.

4.1 Conceptual model

The planning and organisation of the documentation takes up a great part of textbooks and manuals on software documentation. A possible reason is that technical editors are mostly forced to plan, investigate for, and write their documentations already at a point of time when the software product to be documented is still being developed and will therefore constantly change.

New versions comprising an extended scope of functions are also developed for software products already on the market. There are support packages and add-ons, which need to be documented. The fact that coding can be changed easily, therefore also affects the creation of software documentation.

Since the 1970s, the process of software development has become increasingly standardised leading to the emergence of the concept of software engineering. The objective is to engage in software development more systematically and to consequently facilitate the calculation of expenditure, to limit project risks to a minimum and to improve the quality of the final product.

Many companies in the software sector organise the development of software using so-called Software Engineering Process Models. The software development process is hereby organised in clearly defined phases. The results of every individual phase is documented, e.g., via specifications and design documentation. Only the compliance with such Software Engineering Process Models enables the certification in respect to quality assurance, e.g., according to ISO 9000 of the International Organization for Standardization, in the first place. The initial point for Software Engineering Process Models is the expected life cycle of the software, which can be differentiated into purpose/problem, problem analysis, specification, detailed design, implementation, trial, operation and maintenance. The development of software according to these phases is also referred to as documentation-oriented or document-oriented development since each development phase is documented with an individual documentation.

Documents can be differentiated according to whether they are externally or internally accessible. The documents merely authorised for internal access comprise documentation regarding the requirement specifications of the software, the structure and the implementation details. Even though these documents usually partially represent integral parts of the underlying contract of development projects and they are sometimes accessible to the customers before the actual software development is initialised, they are not accounted to be software documentation in a narrow sense and are not part of the scope of delivery of the software product. Internal documents usually also include descriptions of the data model of a software component as well as documents related to data bases and data files. All these documents depict the interior structure of the software in detail. This, for example, also comprises the source code and the comments contained herein.

Technical editors are usually responsible for the external documents. These are also referred to as user documentation. The term as such implies the usual target group of software documentation, even though the user documentation may also comprise documentation intended for system administrators or IT consultants. In most cases the user documentation is required to present the software uniformly and less abstract than development documentations.

The Software Engineering Process Models of the software developers mostly offer several equally footed approaches to the conceptual model of the software which usually disposes of a high level of abstraction and detail.

The central point of attention within the internal documentation of software development is the so-called Conceptual Model. It depicts the basic characteristics of the software. The core objective is to clearly signify the connection of the real world with the internal, software-technical world. The conceptual model normally comprises three approaches: one regarding the data structures, one regarding the functions and one regarding the processes which are mostly depicted in form of diagrams (Hansen and Neumann 2005: 169).

The data related approach (also referred to as the data model) is generally considered to be the focal point of the conceptual model. It determines all technical objects with which the software records the real world and documents the respective interdependencies. The function related approach determines the functions provided by the software. The process related approach connects the data and function approach and presents the internal processes of the software, in particular the interaction between the technical objects.

Like all models, the conceptual model is only able to cover a selection of the real processes and objects. It is possible that one issue may lead to several equally footed conceptual models, even though every development project must finally agree upon exactly one conceptual model.

4.2 Ideals of the software documentation

The authors of textbooks and manuals on software documentation implicitly address the conceptual model of software development. Due to the diversity of software products – reaching from office software over scientific software to computer games – they do however not go into detail. They rather attempt to cover the similarities of the software products and offer strategies for their documentation. What they do have in common is the perception that software consists of functions, is controllable via interfaces, is applied for certain activities and disposes of a complex inner life. The point in which textbooks and manuals differ is the priority which is assigned to these characteristics. In this respect, both a trend towards an orientation to the conceptual model, as well as an orientation reaching as far as the application domain can be found.

Rupietta, in 1987, set the priority on the functions provided by the software. In his opinion, software documentation depicts a descriptive system model, which is consistent and accords with the actual construction of the software product as far as possible. Therefore, documentations should mainly cover the relationship amongst the functional and constructive components. In the course of this, the respectively applied programming model has a great influence on the software documentation. It is utilised as the basis to determine which objects, on the technical level, are to be documented (e.g., programmes, functions, data structures, interfaces, classes and methods) and how these components are combined to a software product either visible or not to a user. As a consequence, the documentation depicts the conceptual model designed by the software.

Also for Price and Korman, in 1993, the functions represent the essential components of software. The authors, however, already differentiate basic functions supporting the user in performing tasks from less important functions. Price and Korman already adopt an external perspective as they require documentation to inform the user more in regard to the application range of software and their work. Only the task oriented functions are relevant for the documentation, but not, however, "the unnecessary bells and ornamental whistles that developers glue on, hoping to distinguish a product from its competition." (Price and Korman 1993: 66).

Bremer, in 1999, focuses on the user interfaces as the central issue of the software documentation. Technical editors are supposed to concentrate on depicting what the user sees and utilises to control the software.

For Barker, in 2003, finally, the actions performed with software are more important than the available functions. The essential characteristic of software is no longer the range of functions or the user interface but the user's actions which the software is integrated in. Therefore software documentation should be focused on the tasks which the software provides support for. The technical editor, in his work, is no longer intended to act according to the functions of the software, nor according to the task-oriented functions, but on the basis of the tasks to be performed by the software as such.

The authors of textbooks and manuals have initially taken up a stance within software. Their priority then was focused on the description of the conceptual model and the provided functions. Such a documentation oriented towards the system depicts the right corner point within the semiotic triangle, i.e., the computer system itself. In more recent publications, in contrast, the authors of textbooks and manuals tend to adopt a standpoint that is located outside the computer system and propose a documentation that is oriented towards the tasks a user performs.

Here, the documentation depicts the connection of the computer system to reality, i.e. how the tasks within the application domain can be mastered with the computer system. In the semiotic triangle this is the connection from the right to the left corner point.

The cognitive perception, the upper corner point, is only indirectly represented in these ideals of software documentation. It is, however, relevant since the correct perception of the computer systems is of importance for a correct use of software.

Price and Korman (1993) respond more clearly to the cognitive perception and adopt the ideal of a convivial documentation (backed by the so-called Tools for Conviviality according to Ivan Illich 1973). In such a documentation, readers will not only be able to find the information they expected, but will also learn to go beyond the intended application of the software and to realise their own visions with it, e.g., to creatively arrange their working environment: "Readers tend to come back to such manuals repeatedly. And instead of just learning how to look up a few items, they branch out and explore, learning more than they first intended. They see more in the product than they did before and understand more about their own work." (Price and Korman 1993: viii). In this case, the computer systems are perceived as media which are transparent and allow the user direct access to reality.

4.3 Interface model

The step from a system-oriented to a task-oriented documentation is based on the implicit assumption of two domains, between which both the software documentation as well as the computer systems act as mediators: the application domain and the system domain.

Rupietta demands that software documentation should depict a descriptive system model that generates an appropriate mental model for the user. His approach is based on the system domain. The basis of task oriented documentation is located in the application domain. The user interface mediates between both domains. This connection works in exactly the same way – it connects the computer system and reality within the semiotic triangle.

According to Bremer, 1999, it is subsequently possible to differentiate between three mental models for software documentation: the mental model of the software developers (expressed by the descriptive system model), the mental model of the user (the model of the application-oriented documentation) and the so-called Interface Model. The Interface Model is developed by the designers of the user interface and realised with the software (see figure 2).

As a consequence of this, contrary to Rupietta and Price and Korman, Bremer primarily directs the focal point of the documentation to the description of the interface model.

According to his perception, the ideal documentation does not shift the mental model of the users on to the level of the mental model of the software developers, respectively on to the level of the descriptive system model, as demanded by Rupietta. The documentation is rather supposed to upraise the model of the user on to the level of the interface model. The interface model is also in the centre of attention of documentation due to the fact that the user interfaces isolate the user from all elements of the technical domain that appear too specialised or not appropriate for the application domain.

Even though suggested by Bremer, the interface model is not only determined by the user interface of the software. The reason for this is that the user interfaces, in regard to their configuration of menus, entry fields and pushbuttons, like all complex sign systems, offers a surplus of interpretable signs. User interfaces contain traces both of the task to be performed by the software as well as traces of the interior of the software. While describing the tasks and functions which are controllable via the user interface and which are necessary for the deployment of the software, the technical editor develops an interpretation of the software against the background of its application. He decides whether the interface model is rather situated on the system domain or further up within the application domain. Subsequently, the role of the software documentation is also to – literally – put down in writing the position of the interface model between the system and application domain.

This role of the software documentation is for instance exemplified in standard business software, which initially requires to be adjusted to the customer's demands after being purchased. This is realised by means of so-called customising activities. Their documentation, due to the numerous possible settings, must naturally contain more information regarding the technical details of the software than instructions for end users. If customers now demand a wider range of possible settings for the customising activities, the software (as referred to in the slang of the software developers) is "drilled down" and changed at the according locations. This procedure also uncovers semiotic traces at the user interface which may cause confusion. Consequently, the documentation must, in turn, present a consistent interface model of the software according to the new conditions.

4.4 Software users

Textbooks and manuals on software documentation may note the fact that software documentation is influenced by the system and application domain

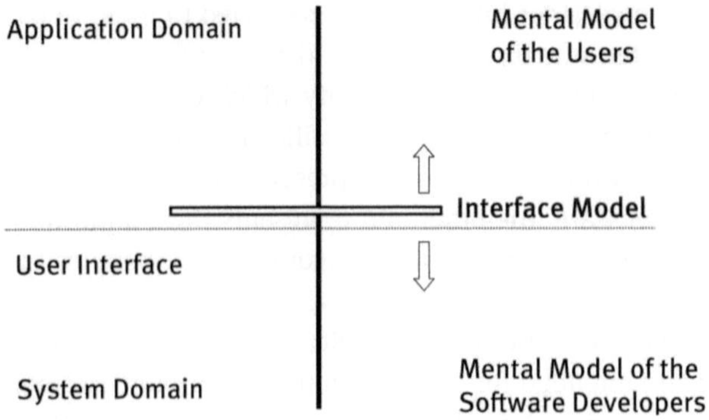

Figure 2: Interface Model

as well as the user interface. The emphasis, however, lies on the target group, i.e. the expected audience of the documentation. The joint perception of all presentations is that the software documentation, in regard to its linguistic form and level of detail, must primarily adjust to the target group, their previous knowledge, reading strategies and attitude to work. The classification of the user and the definition of the target group is therefore considered to be the most important classification criterion for software documentation.

However, a standard target group definition has not yet been developed. One reason for this is that the defined target groups, depending on the type and scope of the utilised categories, can differentiate themselves to a randomly extensive range. Proposed categories are age, current level of training, or the sporadic or frequent use of the software (Mehlenbacher 2003: 531). In practical experience, the creation of individual software documentations for a random number of target groups is not possible as the budget is limited and the delivery channels for documentations are restricted. Subsequently, organisational and not least financial reasons speak against too many target groups. For this reason, textbook and manual literature mostly features general categorisations intended to provide support for the definition of the main target group.

4.4.1 Previous knowledge and desired knowledge

On the one hand, the attempt is made to explain which knowledge the readers are supposed to acquire through the software documentation. The classification criteria comprise the profession of the user, the lines of business like IT departments as well as the management. The management for instance, will only receive the project documentation, the IT department the system documentation and the specialist department the user documentation. Also included is the differentiation whether information is restricted to internal use or if it may also be issued externally (cp. Lehner [1994] 1999: 39).

On the other hand, the attempt is made to clarify which experience and which previous knowledge the readers have. The classification of the previous knowledge uses a scale with two ends. One end represents the knowledge in the application area of the software, the other end represents the skill in handling the software. Between these two poles, are categories as for instance the general knowledge in deploying software in the respective application area (Rupietta 1987: 34–35), or the experience acquired with a comparable software product, or with the software intended for description itself (Price and Korman 1993: 35). The quality of the previous knowledge is usually categorised in three levels which may relate to any of the mentioned knowledge areas. A differentiation is made between beginners or novices and experts or specialists (DIN: VDE 1000, Grünwied 2005: 70). In between are the merely instructed persons (DIN: VDE 1000), i.e. those on an advanced level (Grünwied 2005: 70, Price and Korman 1993: 35) or the experienced users (Barker 2003: 390).

4.4.2 Reading strategy

In the 1990s a task-oriented software documentation was increasingly propagandised which is also reflected in a different type of target group definition. The main emphasis was no longer directed towards the knowledge of the readers but rather towards their reading strategy, i.e. the way they access the documentation (directed reading, skipping, scanning) and for what purpose the documentation is required.

Brockmann (1990: 191) differentiates between the poles Reading-to-learn and Reading-to-do, between which he assumes a continuum of reading strategies. On one hand, the readers want to learn something about the software, and on the other they want to perform certain tasks with it. Within other presentations, complementary to these reading strategies, the writing strategies Writing-to-teach and Writing-to-guide are mentioned, mostly in combination with the third category Writing-to-support, which aims at supporting the users in solving problems with the handling of the software (e.g., Barker 2003: 21).

4.4.3 Attitude towards work

An even more extensive context is focused on by Barker (2003: 15). Neither knowledge nor reading strategies, but the attitude to work with the software is to be reflected by the documentation. Based on sociological and psychological description categories, he differentiates between the default user and the task-oriented user.

The default user, due to the work at the monitor, experiences separation and isolation from his work colleagues and senses the possible control by managers through the networks of the computer. The division of work accelerated by the software is accompanied by abstract tasks for the default user, which leads him to regard his own work, his knowledge and abilities as less valuable.

The task-oriented user is the counter-model to the default user and the role model technical editors are supposed to orient themselves on according to Barker. The task-oriented user senses the work processes, changing due to the software, as a challenge and does not only perform his tasks which were closely narrowed down by the work division, but he also has the desire to look behind the user interface in order to understand the concepts of the software.

4.5 Ideal of a consistent language use

The classification of the users is utilised in textbooks and manuals of software documentation to determine the structure and linguistic design of the software documentation. Software documentation comprises a multitude of texts, which are very different in format and size, including menu items, error messages, function lists up to basic business information. In spite of this, relatively simple rules to the language use can be found in textbook and manual literature.

4.5.1 Tutorial, procedure and reference

Barker (2003: 103), for instance, differentiates the categories of Tutorial, Procedure and Reference. These categories are considered as design principles, which are oriented towards the application scenarios and represent an aid for the structuring of the documentation and the deduction of an appropriate language use. This differentiation can also be found in other manuals.

Tutorials emphasise the conceptual information, e.g., the structure, operation, as well as various different application scenarios of the software. They are supposed to support the reader to learn via the software. This type of documentation also comprises guided tours, demos and getting started guides.

The main focus of the procedures is, above all, directed to the step-by-step instructions as well as navigation and entry aids which support the readers in performing their tasks. Background information regarding the operation mode of the software is rarely included. This type of documentation further involves context-sensitive aids such as roll-over help, field help, but also instruction manuals containing step-by step instructions.

The reference sections comprise the range of functions in a systematic manner. The authors, for instance, provide the user with support in the search for individual adjustments, for functions or interpretation of error messages. The corresponding documentation types are attachments such as FAQs, readme-files, job aids and flipcards.

4.5.2 Rules for language use

The linguistic form of the software documentation is a main topic of textbook and manual literature. Thus, rules for the language use are formulated which are supposed to help optimally design the software documentation in reference to the communication objective. These rules usually cover all linguistic levels reaching from the word to the text level and, e.g., address both the terminology as well as the standard wording of sentences, structural proposals regarding paragraphing, listing as well as the combination of text components to form an extensive software documentation.

The rules contain general recommendations in regard to style, e.g., the avoidance of the use of unusual words, the use of short sentences, the avoidance of participle and nominal phrases. Other rules reflect the results of studies on the topic of the understandability of text (e.g., Lehner [1994] 1999: 94; general information on the optimisation of the technical editorial office c.p. Göpferich 1998a). In addition, there are also company-specific standards and guidelines.

The main concern of the authors of guidebook and manual literature is a consistent language use, both in respect to the expertise (e.g., Bremer 1999: 22) as well as in respect of the phrasing. Equal circumstances should be equally depicted and different ones respectively in different ways (e.g., Barker 2003: 388; Rupietta 1987: 85).

Terms for technical objects should be used well-defined. Headings, references and listings should be uniformly formulated, if possible, using the same sentence construction (e.g., Barker 2003: 388). This parallelism on the sentence level is continued for macro structures such as paragraphs that contain information on objects, functions or procedures. Those paragraphs should also be formulated according to standardised schemes (e.g., Rupietta 1987: 85).

The description formulas mostly comprise several paragraphs with corresponding textually clearly defined contents, the order of which is founded on the structure, appearance or the handling of the software intended for description. In case of procedures, the task should initially be depicted, followed by the actions necessary for their respective performance (Price and Korman 1993: 362); following the description of what the user is supposed to do, the system response should follow (Barker 2003: 80). This should initially comprise a description of what the system does followed by an explanation of the operation modus (Barker 2003: 54).

Company-specific standards additionally require a certain order in which such background information is made available (Schmidt 2005). In case of procedures, these should first state the corresponding technical prerequisites and subsequently the steps to be performed by the user.

5 Perspectives

Textbook and manual literature on software documentation and publications from the field of semiotics as well as cultural studies have little in common at first glance. Cultural studies refer to the handling of computers as a cultural technique but do not respond to the documentation problem. This however, is of central significance within literature on software documentation.

The semiotic approach, in contrary, emphasises more clearly, why computer systems are particular machines: that they are practically applicable within all sectors of life due to their abstractness that they are both tools as well as media and initiate numerous sign processes, support the user in perceiving the computer system and reality, but also that they can be the cause for a multitude of misunderstandings.

Literature on software documentation merely implicitly responds to these general findings. Its emphasis is directed towards the craft of the technical editors. As a consequence, textbooks and manuals primarily depict the organisation of documentation projects and make suggestions as to the design and phrasing of documentations.

However, a development is noticeable in textbooks and manuals, which can be aligned with the semiotic approach. In the 1980s, the system domain was still the main focus of the documentation. The ideal was represented in a descriptive system model. In the 1990s, the focal point shifted towards the application domain. At that point of time, the task-oriented documentation was stated as the ideal. Soon afterwards, the convival documentation became a further ideal.

Within the semiotic triangle, these ideals can be allocated to respectively different corner points. The main emphasis of the descriptive system model is directed towards the description of the computer system as a sign system. The focal point of the task-oriented documentation is the reality outside of the computer system. It primarily regards computer systems as tools, which the user desires to use to perform certain tasks. In case of the convivial documentation, the main focus is directed towards the mental perception the user develops of computer systems and the non-verbal reality.

The fact that these three approaches are closely linked is elucidated by the interface model. On the one hand, the user controls the computer via the interface and applies it as a tool. On the other hand, he perceives the reality through the computer system via the interface and develops his own perception of reality, however also via the image of the reality as presented by the computer system. The interface model itself is jointly depicted by the user interface and software documentation and is subsequently determined.

The semiotic triangle and the interface model are, at first glance, still quite general and far away from the actual practice of technical editors. Both models however, feature perspectives which can be utilised for the training of technical editors.

Thus, the semiotic approach connects the ideals of the descriptive system model and the task-oriented documentation. This makes apparent that both the analysis of the application domain and an independent perception of the technical aspect of the computer systems are necessary for good software documentation. Within practical experience, this independent perception of software is often nonexistent. Many technical editors orient themselves on the system domain. The reason for this is the development process and the fact that software can be changed easily. This software reveals many semiotic traces in the course of the development process forcing technical editors to look at the technical context behind the user interface, and to subsequently also depict the individual technical objects such as functions, fields and structures in the documentation. However, a uniform point of view based on definite criteria is not to be found. The interface model could – in the sense of the semiotic meta-program – reveal design principles with which software could be depicted with the desired expertise and consistency, without exclusively utilising artefacts, such as technical development objects or elements of the user interface, for this purpose.

In addition, the semiotic approach connects the task-oriented documentation with the convivial documentation and therefore expands the objective of software documentation. It is not only supposed to aid the user in performing tasks with computer systems, but also to support them to gain more insight into the application domain by means of the computer. The convivial documentation finally offers the user an opportunity to also use computer systems as a medium and to deploy them creatively and to utilise them as a cognitive instrument. This leads to the desire and the objective, not only to write an accurate and useful, readable and comprehensive software documentation, but at the same time, also a documentation that is interesting and inspiring to the reader.

References

Andersen, Peter Bøgh
1997 *A Theory of Computer Semiotics. Semiotic approaches to construction and assessment of computer systems*. (Cambridge Series on Human-Computer Interaction; 3). Updated Edition, Cambridge: University Press.

Andersen, Peter Bøgh, Hasle, Per and Aage Brandt, Per
1997 Machine semiosis. In: Posner, Roland, Robering, Klaus and Sebeok, Thomas A. (eds.), *Semiotik. Ein Handbuch zu den zeichentheoretischen Grundlagen von Natur und Kultur*. 1. Teilband, 548–571. (HSK 13.1) Berlin/New York: de Gruyter.

Barker, Thomas T.
2003 *Writing Software Documentation. A Task-Oriented Approach*. (Ally and Bacon series in technical communication). Second Edition. New York et. al.: Longman.

Borwein, Jonathan, Bailey, David and Girgensohn, Roland
2004 *Experimentation in Mathematics. Computational Paths to Discovery*, Natick, MA: A K Peters.

Bremer, Michael
1999 *The User Manual. How to Research, Write, Test, Edit and Produce a Software Manual*. Concord, CA: UnTechnical Press.

Brockmann, John R.
1990 *Writing Better Computer User Documentation: From Paper to Hypertext*. New York, NY: John Wiley & Sons.

DIN 66230
1981 *Information processing; software documentation*. Berlin: Beuth.

DIN VDE 1000
2011 *General principles for the safe design of products*. Berlin: Beuth.

Göpferich, Susanne
1998a Möglichkeiten der Optimierung von Fachtexten. In: Hoffmann, Lothar, Kalverkämper, Hartwig and Ernst Wiegand, Herbert in Verbindung mit Galinski, Christian und Hüllen, Werner (eds.), *Fachsprachen: Ein internationales Handbuch zur Fachsprachenforschung und Terminologiewissenschaft*. New York/Berlin: de Gruyter: 888–899.

Göpferich, Susanne
1998b *Interkulturelles Technical Writing. Fachliches adressatengerecht vermitteln. Ein Lehr- und Arbeitsbuch*. Tübingen: Narr.

Grünwied, Gertrud
2005 *Software-Dokumentation. Grundlagen – Praxis – Lösungen*. (Kontakt & Studium 668.) Renningen: Expert.

Hansen, Hans Robert and Neumann, Gustaf
2005 *Wirtschaftsinformatik 2. Informationstechnik*. 499 Abbildungen, 9., neu bearbeitete Auflage. Stuttgart: Lucius & Lucius.

Horton, William
1994 *Designing and Writing Online Documentation. Hypermedia for Self-Supporting Products.* Second Edition. New York: Wiley.
Illich, Ivan
1973 *Tools for Conviviality.* New York: Harper & Row.
ISO 9000
2005 *Quality management systems – Fundamentals and vocabulary.* Berlin: Beuth.
Kilian, Crawford
2007 *Writing for the Web.* Complete Edition. 3. Auflage. Brampton, Canada: Self-Counsel Press.
Krämer, Sybille
1988 *Symbolische Maschinen. Die Idee der Formalisierung in geschichtlichem Abriss.* Darmstadt: Wissenschaftliche Buchgesellschaft.
Krämer, Sybille
1998 Das Medium als Spur und als Apparat, In: Krämer, Sybille (ed.), *Mensch – Computer – Realität. Wirklichkeitsvorstellungen und Neue Medien,* 73–94. Frankfurt am Main: Suhrkamp.
Krämer, Sybille and Bredekamp, Horst
2003 Kultur, Technik, Kulturtechnik: Wider die Diskursivierung der Kultur. In: Krämer, Sybille and Bredekamp, Horst (eds.), *Bild, Schrift, Zahl,* 11–22. (Reihe Kulturtechnik) München: Wilhelm Fink Verlag.
Lehner, Franz
[1994] 1999 *Software-Dokumentation und Messung der Dokumentationsqualität.* Mit 55 Abbildungen. Berlin: Logos-Verlag.
Mehlenbacher, Brad
2003 Documentation: Not yet implemented, but coming soon! In: Jacko, Julie A. and Sears, Andrew (eds.), *The Human-Computer Interaction Handbook. Fundamentals, Evolving Technologies and Emerging Applications,* 527–541. Mahwah/New Jersey: Lawrance Erlbaum Associates.
Nadin, Mihai
1998 Computer. In: Bouissac, Paul (ed.), *The Encyclopedia of Semiotics*, 136–138. New York/Oxford: Oxford University Press.
Ogden, Charles Kay and Richards, I. A.
[1923] 1989 *The Meaning of Meaning.* Frankfurt am Main: Suhrkamp.
Price, Jonathan and Korman, Henry
1993 *How to Communicate Technical Information. A Handbook of Software and Hardware Documentation.* Redwood City: Benjamin/Cummings Publishing Company.
Rupietta, Walter
1987 *Benutzerdokumentation für Softwareprodukte.* (Angewandte Informatik Band 3.). Mannheim/Wien/Zürich: BI-Wissenschaftsverlag.
Schmidt, Vasco Alexander
2005 Technisches Schreiben bei SAP. Dokumentation betriebswirtschaftlicher Standardsoftware. In: Jakobs, Eva-Maria et. al. (eds.), *Schreiben im Beruf,* 73–92. Frankfurt am Main: Peter Lang.

Schmidt, Vasco Alexander
2007 Inneres und äußeres Arbeiten am Text. Das Verhältnis vom Schreiben und Redigieren zum Sprachsystem in den Kommunikationsbereichen Zeitungsjournalismus und Softwaredokumentation. In: Wich-Reif, Claudia (ed.), *Strukturen und Funktionen in Gegenwart und Geschichte. Festschrift für Franz Simmler zum 65. Geburtstag*, 629–652. Berlin: Weidler.

Marc Hermeking

Cultural aspects of technology and documentation: Contributions from the field of intercultural communication

1 Introduction

"The idea of a 'pure' technology is nonsense" – this marked statement resulted from sociologic research on the development and construction of various technical products of the past (Bijker and Law 1992: 3). However, it is not only the process of creation or production but the consumption or usage of 'hard' technology which is influenced by 'soft' social and, even more important, cultural factors: An increasing number of empirical studies and examples confirm the manifold and sometimes subtle, but nevertheless very critical influences of culture on technology. Consequently, and unlike in some international marketing literature, technology in general should not be taken simply for to-be-standardised-easily 'culture-free' products that, additionally, would support the development of an increasingly homogenised global 'one-world' culture (Hermeking 2008).

The following text provides some empirical examples of the influence of culture on the usage of technology and technical documentation, which will be illustrated with some relevant theories and models from the field of intercultural communication.

2 Intercultural communication and technology

The implications of cultural diversity and the interactions between members from different cultures are central issues of the academic and practical field of intercultural communication, which is based on various academic disciplines including, as the most essential ones, cultural anthropology/ethnology, cross-cultural psychology, and translation studies/linguistics. Though personal face-to-face interaction is its traditional focus, intercultural communication is related in many ways to material culture in general (Roth 2001) as well as to technology and its cross-cultural transfer in particular (Hermeking 2001). The pioneer work of US-American anthropologist Edward T. Hall of the 1950ies is commonly regarded as the starting point of this discipline – in the meantime, intercultural communication has found broad acceptance and additional theoretical development by many others.

The work of Edward T. Hall, who died in 2009, was influenced by anthropological theories of Functionalism, according to which culture comprises material and immaterial products as instruments to cope with problems which all human beings are exposed to during their existence. These problems are identical worldwide, but which instrument out of a variety is preferred depends on the specific priorities or values that people in social units or groups, i.e. cultures, collectively share. Consequently, different groups or cultures are characterised by different sets of values which guide their existence. Technical products or instruments, as such, are clearly included in that conception of culture.

Hall's work focused on the interdependence of culture and communication, which is also expressed in his most central statement that "culture is communication". In accordance with that, he introduced ten universal cultural categories as "Primary Message Systems", of which three categories – the orientation towards "Temporality" (time) as well as "Territoriality" (space), and "Interaction" (context and message flow) – were elaborated by Hall in more detail and consequently gained larger popularity in the following years (Hall 1959; Hall and Hall 1990). Being similar to other popular categories like Hofstede's (1991) cultural dimensions, Hall's model is based on macro-cultural entities like lingual communities, regional populations, or even nations, which has both advantages and disadvantages (Hermeking 2005).

Though the cultural categories of time, space and context/message flow were originally not explicitly related to technology and technical documentation, they can well be applied to that matter. This, in Hall's words, may meet "the classic goal of true science, which is to explain more and more events with fewer and fewer theories" (Hall 1959: 93). The very essential reason, however, is that communication is an increasingly central element of both the usage of modern technology and of technical documentation.

Around the world, nearly all kinds of technical products are characterised by an increasing computerisation and an increasing share of elements of information and communication technology (ICT), so the handling of modern technology is to a large part a process of communication/interaction between human beings and machines. Consequently, distinctions between the usage of technology in general, of ICT in particular, and their related technical documentations are decreasing while the (inter-)cultural aspects of computer mediated communication, human-computer interaction, and digital user interfaces are gaining increasing importance. This will be demonstrated by the following examples.

3 Culture and technology usage: The internet

A prominent example for the global usage of technology, or ICT in particular, is the internet and its most popular user interface, the World Wide Web. However, worldwide consumption data over the last decade show remarkable national differences in the numbers of (active) internet users (Hermeking 2005, 2007). Although those figures change continuously over time, there is a clear continuum of descent from high internet usage in the (developed, western) north to low usage in the (often less developed, non-western) south – also known as the 'Digital Divide'. Economic development, like technical infrastructure or national income per capita, is only one (hard) factor. A close look at the different usage rates in highly developed countries like the USA, Sweden, France or Japan, for example, reveals that some cultural (soft) factors play an even more important role.

Utilising Hofstede's (1991) model, the influence of cultural values on internet usage can be visualised by the significant correlation of many countries' usage data with the country scores of the two cultural dimensions "individualism" and "uncertainty avoidance" (Hermeking 2005, 2007). Like all artefacts, the internet is a specific cultural product which originates from a specific cultural environment. Accordingly, the diffusion of the internet from its country of origin, the USA, to other countries of the world seems to depend – among other factors – on their cultural similarity to the USA, specifically with respect to cultural values of strong individualism and low uncertainty avoidance. The latter represents a cultural openness to adopt innovations like the 'new medium'; the former corresponds to this relatively impersonal medium of communication between individuals.

Hofstede (1991) refers to Hall's model of context and concludes that high individualism is very often connected to "low-context" communication whereas low individualism (collectivism) is very often connected to "high-context" communication. Members of low-context cultures have less personal contact with each other; therefore communication must be very detailed and very explicit. A big amount of formal information is communicated in a direct way, often by way of written texts ('high content'). Members of low-context cultures are, for example, Scandinavians, Germans, Anglo-Americans and Britons. In contrast, members of high-context cultures have closer and more familiar contacts with each other; a lot of information is already shared among them, and therefore their preferred mode of communication is more informal, indirect and often based merely on symbols or pictures ('low content'). Members of high-context cultures are, for example, the French, Spanish, Latin-Americans, Arabs and the Japanese (Hall 1976; Hall and Hall 1990)

According to Hall's context model and the globally different internet usage rates, this new medium in general is more preferred in low-context cultures than in high-context cultures – with few exceptions, of course. This obvious impact of cultural communication preferences on global internet usage shows interesting analogies with the global consumption pattern of traditional media such as newspapers and TV (Hermeking 2005, 2007). In general, global traditional media consumption reveals a consistent tendency of a higher popularity of print media in low-context cultures, in contrast to a higher popularity of TV in high-context cultures. This quite evidently corresponds with the culturally preferred communication styles, as TV is mostly based on pictures and sounds whereas print media are more or less based on explicit, written texts.

The analogies with print media consumption indicate the fact that the nature of the internet during the last decade has been more similar to print media, mainly based on a text-heavy, explicit and informational communication style, which is therefore accepted most of all in low-context cultures like the culture of origin, the USA, where the internet originally was developed for explicit informative and scientific reasons (Arpanet). According to international consumer research, frequent internet users worldwide tend to read more books and newspapers than non-users, so they may be characterised as a low-context "information elite" (Hermeking 2005, 2007). Recently, however, as a result of more efficient data transfer and mobile usage, the internet has gained much popularity in some highly developed centres of high-context Southeast-Asia, for example, while the newer internet (Web 2.0) offers much more pictures, videos and sounds (for example, YouTube, or IPTV). This might indicate a gradual change of that medium towards high-context communication preferences, which is also reflected in its global usage.

The World Wide Web is characterised by a higher degree of variability of its communication style than traditional media – elements of both print media and TV (and telephone, etc.) can be combined here. Although the internet originally was inclined strongly towards low-context communication, cross-cultural comparisons of many (commercial) web sites reveal cultural differences and adaptations of their design/style, which show remarkable analogies to advertising in the traditional media with respect to culture-specific advertising styles and creative strategies. In general, indirect and transformational advertising messages creating emotions through pictures and entertainment are more favoured in high-context cultures (like France or Japan) whereas direct and rational advertising messages providing product information above all play a more important role in low-context cultures (such as Germany or many parts of the USA). This general tendency – with some exceptions, of course – also holds true for the design or communication style of web sites along several structural design criteria, which, in addition, also correspond to culture-specific orientations towards time (Hermeking 2005, 2007).

According to Hall, low-context communication often correlates with a "monochronic", and high-context communication with a "polychronic" time orientation (Hall and Hall 1990). In monochronic cultures, time is linear, passing ('time is money'), handled precisely, and divided in an orderly fashion. Thus, action chains are generally compartmentalised and sequential; one thing is completed after the other. In polychronic cultures, time is generally regarded as circular and repeating, subject to social relationships or needs and therefore handled in a flexible, imprecise way. Action chains are structured in a less detailed way and interrupted more often since many things are done simultaneously.

Correspondingly, low-context style web sites that carry a great deal of explicit informational content are detailed, compartmentalised, structured in an orderly way, and also supported by many explicit navigation tools such as sitemaps, index registers, and search engines to facilitate a quick orientation and a time-saving site use. In addition, missing multimedia presentation and a text-heavy layout of the naturally voluminous web sites reduce download times and correspond to a monochronic time orientation. In contrast, a much higher ratio of entertaining visuals, animated illustrations, and even real multimedia elements (sound, video) in high-context style web sites certainly increases download time, but this does not matter so much. Here, explicit navigation support is rare since neither a strictly ordered route through the less detailed site structure nor very quick orientation is necessary. Implicit symbolic cues, however, support an intuitive navigation, which enhances both entertainment appeals and positive affection in web site use (Hermeking 2005: 207).

Of course, many additional culture-specific design criteria of web sites like visual motifs, symbolic colours, styles of address and tone, etc., could be mentioned here. In addition, it is important to note that there are some exceptions from these general tendencies. These are either based on diverse non-cultural contingencies like the represented branch or product type, or they may be subject to some additional cultural conditions like a traditionally more visual and symbolic nature of written text and a specific rationale towards 'both text and visuals' in Southeast-Asia (Hermeking 2007, 2008).

Both the usage of the internet and the design of web sites, as elements of its most important user interface, are prominent examples for the close interconnections between technology, communication and culture. Another aspect of that interconnection is the internet's central role as a new medium of technical documentation. Information about technical products and instructions for its usage, which were traditionally printed in operation manuals and the like, are increasingly distributed via the web as digitalised documents for download or for presentation on screen. Such technical documentation via another technology may be characterised as a new, digitalised form of offline instructions, which as such also reveal many influences of culture.

4 Culture and technical documentation: Offline instructions

Being a specific form of communication, technical documentation reflects both cultural communication styles and the culture-specific usage of technology. Traditionally printed documentation with instructions describing technical usage and maintenance offline, i.e., not necessarily during usage of the related technology, is strongly affected by such cultural influences. This, for example, was also demonstrated by the case of a medium-sized hydro-electric power plant in Colombia (Hermeking 2001).

The turbine shafts of the power plant had to be maintained regularly by new oiling 'every 2,500 hours of operation', which, however, was repeatedly not carried out by the locals. This led to technical problems including turbine breakdowns and power failures, which could put public acceptance of the complete electrification project on stake. In addition to explicit instructions in their detailed and voluminous Spanish-translated operation manuals, the German supplier therefore installed both chronometers counting the hours of operation and big plates with maintenance instructions right above the turbines. All this did not help, however, and led to technical problems.

There seemed to be no solution until a culture-sensitive expert advised the German supplier to handle the matter in a very different way. The prescribed 2,500 hours were equivalent to 15 weeks or nearly four months, so the maintenance procedure had to be carried out three times a year. Three Christian festivals following each other at a period of about four months were finally chosen and the maintenance date was fixed right according to these days. To the German supplier's surprise, there was no further maintenance problem.

Some trouble could have been avoided if the German supplier as a member of a low-context culture had adapted the technical system from the very beginning to a high-context culture which, in this case, is less literal or print-oriented and therefore less fond of reading many manuals, instructions or measuring-devices and which is, additionally, less abstract and future time-oriented, but more present and concrete time-oriented than German culture (Hall and Hall 1990). As members of a monochronic culture, Germans divide time precisely into hours, minutes and seconds; by doing so, time is handled as an abstract and linear mathematical figure. This may confuse members of a polychronic culture like the Colombians who handle time in a more flexible or imprecise way. For them, concrete time is more circular and repeating, like natural seasons.

Ignoring the cultural backgrounds of the different relevance of low-context vs. high-context style product documentation may therefore be seen as one important reason for many problems of maintenance, operation and the international transfer of technological know-how in general. But culture-specific styles of communication and perception do not only influence the literal vs. visual aspects of technical documentation. Even technical drawings, sketches and pictures are interpreted differently between different cultures, which, for example, were demonstrated by the case of a mine in South-Africa (Hermeking 2000).

To reduce accidents with mining machinery during technical operations carried out by the mostly illiterate African mineworkers from less developed regions, warning signs with abstract pictures (pictograms) had been installed by the US-American mine management. The desired effect was not achieved, however, and a high accident rate remained. Finally, the mine management started to investigate why the signs and pictures were ignored. The reason was that the local workers perceived the abstract pictures in a two-dimensional way, which led to a totally different interpretation of the messages than the western side with its habit of a three-dimensional pictorial perception originally intended. Thus, the abstract picture of a worker, for example, carrying a long pole that injures the head of another 'smaller' worker far behind him was not perceived as the intended warning of head injuries, but as a nice scenario of 'a man accompanied by a child in the sunshine'.

As members of a strong high-context culture, the African mineworkers were used to concrete, full-colour visual information, which could be given by videos or photographies; those visuals are rich of symbolic cues and allow contextual interpretations of the represented scenario. In contrast to that, the western style of very abstract, black and white visual schemes with few contextual cues, and the specific tradition of three-dimensional expressions did not meet their habits of communication and perception.

Of course, culture-specific traditions of visual perception are not restricted to the non-western or less developed world – they also exist among highly developed countries. An example for that are the different standards of projections in technical drawings like the "Third Angle Projection" (ISO method 3) in the USA vs. "First Angle Projection" (ISO method 1) in continental Europe (Schmitt 1999). Accordingly, the visual arrangements of technical parts 'from top, bottom, right, and left' may be interpreted in the completely opposite way.

The following example from highly developed countries in Southeast-Asia also demonstrates that even among visual technical documentation there are many cultural differences, which, in addition to Hall's context model, are also related to other complementary models from the field of intercultural communication (Hermeking 2008).

Comparisons of technical documentation of nearly all kinds of products reveal a much higher ratio of pictures and visuals in Southeast-Asian countries like Japan, China, or South-Korea than in western countries like Germany, where text-heavy contents reveal a stronger division into separate chapters. However, the culture-specific visual style of the former documentation is even more significant. Unlike in western countries, most technical products are not represented directly by 'realistic' photographies or precise technical drawings, but more indirectly by comic-like illustrations. Like 'living' little characters in a cartoon, most technical products are drawn with eyes, a mouth, hands and feet, which make these products gesticulate by way of non-verbal communication or express themselves verbally with text bubbles. In addition, such documentation is also characterised by many natural motifs like flowers, birds, butterflies, etc. In sum, from a western perspective, all this may seem to be naive or even more 'childish' than 'technical'.

Again, the communication preferences of both high-context cultures and low-context cultures are represented in the layout of documentations. This also holds true for the stronger divisions or compartmentalisations of the detailed literal contents in western countries like Germany, where, in addition, action chains are more orderly (i.e., 'step-by-step') structured in a monochronic way. Even the more indirect representation of technical products by playful illustrations or quite imprecise drawings in contrast to real photographies refers to high-context preferences.

In addition to that and maybe even more important, the significant visual style of illustrating technical objects like nice little creatures combined with many natural motifs is related to aspects of culture-specific attitudes towards technology, which are based on a culture-specific style of thinking. The latter, in contrast to a western 'either-or' logic, is much more characterised by a specific Southeast-Asian 'both-and' logic, which has been described, for example, by Johan Galtung (1981), Richard Nisbett (2003), and others (Hermeking 2008).

According to such a style of thinking, there is a lot of harmonic co-existence of entities that, from a western point of view, seem to be contradictory and excluding each other – like multiplex religious adherences, traditionalism (for example, Feng Shui) and modernity in everyday life, both centralised and market economies (in China), ambivalence in verbal communication (both 'yes' and 'no'), a high ratio of text in visual heavy layout of web sites, advertising and technical documentations, and much more. Such a concept of harmony is also represented in the Taoist symbol of the opposing elements Yin-Yang ('shadow/sunlight'). Corresponding with that, even technology and nature are by no way contradictions, but interdependent entities that are combined in harmony – nature is to be perfectionised by technology and technology is to be perfectionised by nature. Technical products consequently are illustrated like living natural objects or friendly little creatures, which are also shown in harmony with other surrounding elements of nature.

The herewith related aspect of a significant enthusiasm towards artificial intelligence and virtual technology in the latter culture refers to another development in technical documentation. Like personal computers, more and more technical products provide information about themselves and operational instructions right during their usage. The cultural influences on that interactive technology are worth being considered.

5 Culture and man-machine interaction: Online instructions

Combined with an increasing share of elements of ICT in nearly all kinds of technical products, the usage of technology is increasingly characterised by online instructions and information given by these technical products or their digital user interfaces during usage. The advantages may be some more comfort and more safety for the users of these products. Problems, however, result from different communication styles of the technical product (or of its producers) and the user's side, which may lead to bad usability, misunderstandings and even serious mistakes in the interaction between the user and the machine. According to a previous long-term survey, about 98% of nearly 1,000 lethal accidents during the use of computerised technology were caused by misunderstandings and mistakes in communication of human-computer interaction (MacKenzie 1996). In a cross-cultural context, differences in cultural communication styles can make these problems even worse, which, for example, was demonstrated by the airplane collision of Ueberlingen (Hermeking 2008).

In 2002, two airplanes collided above Ueberlingen, Germany in a nearly rectangular way. The two airplanes – a Russian machine on its way from Moscow to Barcelona and a German machine on its way from Bergamo to Brussels – both had an identical height of 36,000 ft. (11,000 m). In accordance with international aviation standards, both airplanes were equipped with a traffic alert and collision avoidance system (TCAS), which calculates in advance the movements of the own machine and of surrounding machines in order to warn or to give system-coordinated flight instructions to the respective pilots. Right on time the TCAS reacted; it instructed both the German pilots to decrease ('Descend!') and the Russian pilots to increase height ('Climb!'). At the same time the regional sky guide instructed one machine to decrease height, but erroneously the Russian machine. While the German pilots were following TCAS' instructions and were decreasing their height, the Russian pilots were confronted with two contradictory instructions – and decided to follow the sky guide's instructions so they were descending, too. Both machines kept an identical height in contrast to TCAS's instructions. A few seconds after the German pilots finally contacted the sky guide, the two machines collided.

The critical behaviour of the Russian pilots can be characterised with their preference for the sky guide instead of the technology, whereas the critical behaviour of the German pilots can be characterised with their preference for the technology instead of the sky guide (who could have been contacted earlier). Ignoring any individual behavioural tendencies, this preference pattern obviously corresponds with a culture-specific tendency of the general attitude towards technology, which in a former study on technology usage in Russia (and other countries) as well as Germany has been summed up as "people are preferred to technology" with respect to the former, as well as "technology is equal to people" (Hermeking 2001: 199, author's translation) with respect to the latter – here, technology sometimes is even clearly preferred to people, like in this case.

Utilising Hofstede's (1991) model of cultural dimensions, this tendency is also related to a stronger orientation towards objective tasks in more individualistic cultures like Germany, in contrast to a stronger orientation towards social relationships in less individualistic (more collectivistic) cultures like Russia. In addition, Hofstede's reference to the correlation of individualism/low-context and collectivism/high-context also unveils a surprising aspect: Hall describes explicit low-context messages as a kind of "digital", and informal high-context messages as a kind of "analogic" communication (Hall 1976; Hall and Hall 1990). Although these terms were originally used as metaphors only, they seem to be quite precise descriptions for the cultural preferences towards online instructions given by human beings vs. digital technology or machines.

From such an intercultural point of view, the pilots of both the Russian and the German machine reacted in a plausible way, i.e., in accordance with their culturally different preferences of communication and technology usage.

These differences collided in an unconscious way with each other via interactive technology. Like the internet, such technology is not culturally neutral either, but tends to represent cultural standards or values and communication preferences of its country of origin.

As a consequence of the collision of Ueberlingen, international aviation standards now prescribe an absolute priority of the machine, i.e., TCAS, no matter what any sky guide may communicate. With respect to unconscious cultural tendencies of behaviour, it is doubtful if such a superficially imposed universal standard of technology usage and communication is to be the one-and-only solution. In Hall's (1959: 93) words: "It is doubtful that anyone ever really changes culture in the sense this term is ordinarily used." Consequently, with respect to cultural communication preferences, a technical adaptation of TCAS' digital user interface and its mode of online instructions towards some more high-context communication style seem to be desirable as well.

Of course, not only in the aftermaths of examinations and reflections of such "moments of catastrophic failure" (Bijker and Law 1992: 2) the development and construction of technology as well as technical documentation and communication design can be, and should be, inspired by contributions from the field of intercultural communication.

6 Résumé

Resulting from an increasing computerisation or share of elements of ICT in nearly all kinds of technology around the world, aspects of communication originally related to the former are becoming increasingly relevant for the latter. Consequently, and also much in accordance with Hall's concept of culture as communication, the culture-specific usage of technology – sometimes also referred to as 'technological styles' or 'technology cultures' – can be described to a large part by culture-specific communication preferences or styles of communication. Similarly, technical documentation and operational instructions, which can be differentiated into traditional offline instructions and online instructions given by an increasingly popular interactive technology, reflect both culture-specific styles of communication and the culture-specific usage of technology.

In the many examples for culture-specific usage of technology and technical documentation, Hall's low-/high-context model, which is most explicitly related to communication, was proved to be very useful and compatible with other relevant and also important macro-cultural categories and models from the field of intercultural communication.

By utilising such models, which by definition are always simplifications of a very complex reality, this discipline does not provide any concise or deep analysis of culture, but a first basic insight into the role and influence of culture on various phenomena of everyday life (including technology) as well as a deeper sensibilisation for that matter. By doing so, intercultural communication does not only improve the mutual understanding as well as the interaction between members from different (national) cultures – it can also act as a mediator between the often universalistic, culture-insensitive disciplines or occupations of technical engineering and the diverse, often technology-insensitive cultural sciences. With their experience in using ethnographic methods, the latter are able to unveil and to explain the essential micro-cultural details in specific situations. By contributions from intercultural communication, members of both of these knowledge-cultures may become motivated and qualified for a prosperous cooperation.

Such cooperation is already on its way. During the last years, several renowned industrial producers of ICT (like Intel, Xerox, Philips, etc.) have started to hire cultural anthropologists in order to gain deeper insights into the culture-specific demands and the forms of usage of their technical products (Squires and Byrne 2002). Beyond the design and construction of technical products for global usage, their appropriate and culture-sensitive documentation is another essential aspect of such investigation.

References

Bijker, Wiebe E. and Law, John
1992 General Introduction. In: Bijker, Wiebe E. and Law, John (eds.), *Shaping Technology/Building Society. Studies in Sociotechnical Change*, 1–16. Cambridge, MA/London: MIT Press.

Galtung, Johan
1981 Structure, Culture and Intellectual Style: An Essay Comparing Saxonic, Teutonic, Gallic and Nipponic Approaches. *Social Science Information* 20 (6), 817–856.

Hall, Edwad T.
1959 *The Silent Language*. Garden City, NY: Anchor Press/Doubleday.

Hall, Edwad T.
1976 *Beyond Culture*. Garden City, NY: Anchor Press/Doubleday.

Hall, Edward T. and Hall, Mildred R.
1990 *Understanding Cultural Differences: Keys to Success in West Germany, France and the United States*. Yarmouth, ME: Intercultural Press.

Hermeking, Marc
2000 The Cultural Influence on International Product Development. *Global Player. International Management and Business Culture* 7 (5): 8–13.

Hermeking, Marc
2001 *Kulturen und Technik: Techniktransfer als Arbeitsfeld der Interkulturellen Kommunikation. Beispiele aus der arabischen, russischen und lateinamerikanischen Region*. Münster/München/New York: Waxmann.
Hermeking, Marc
2005 Culture and Internet Consumption: Contributions from Cross-cultural Marketing and Advertising Research. *Journal of Computer Mediated Communication* 11 (1), article 10, 192–216. Retrieved 2009 July, from http://jcmc.indiana.edu/vol11/issue1/hermeking.html.
Hermeking, Marc
2007 Global Internet Usage, Web Site Design, and Cultural Communication Preferences: Contributions from Cross-cultural Marketing and Advertising Research. In: St, Kirk.Amant (ed.), *Linguistic and Cultural Online Communication Issues in the Global Age*, 160–176. Hershey, PA: Information Science Reference.
Hermeking, Marc
2008 Kulturelle Kommunikationsstile in der Mensch-Maschine-Interaktion: Einflüsse auf technische Bedienungsanleitungen und Internet-Webseiten. In: Rösch, Olga (ed.), *Technik und Kultur*, 163–185. Berlin: News & Media.
Hofstede, Geert
1991 *Cultures and Organisations: Software of the Mind*. London: McGraw-Hill.
MacKenzie, Donald A.
1996 *Knowing Machines: Essays on Technical Change*. Cambridge, MA/London: MIT Press.
Nisbett, Richard E.
2003 *The Geography of Thought: How Asians and Westerners Think Differently ... and Why*. New York, NY: The Free Press.
Roth, Klaus
2001 Material Culture and Intercultural Communication. *International Journal of Intercultural Relations* 25 (5): 563–580.
Schmitt, Peter A.
1999 *Translation und Technik*. Tübingen: Stauffenburg.
Squires, Susan and Byrne, Bryan (eds.)
2002 *Creating Breakthrough Ideas: The Collaboration of Anthropologists and Designers in the Product Development Industry*. Westport, CT/London: Bergin & Garvey.

Gerhard Banse and Andreas Metzner-Szigeth

Cultural diversity and new media – their interaction as an element of European integration: Elaborating a European research network

1 Introduction

The paper first summarises the background of the genesis and the subsequent development of the European research network "cultural diversity and new media" (CULTMEDIA). Then it gives an introduction into the leading questions and basic approaches of its research programme. The three main strategic objectives addressed by CULTMEDIA are 1.) to update the scientific debate concerning the relationship of culture and media, 2.) to research the interactions between cultural diversity and new media, and 3.) to discuss their role as an element of European integration. They are subsequently tackled in the following chapters which outline how CULTMEDIA's research activities could be – and, meanwhile, after some years of network activities, have been elaborated on three levels, the "abstract-conceptional", the "discursive-comparative", and the "synthetic-conciliar" level of research. Chapter 4 deals above all with the notions of culture, technology and communication in order to establish an analytical framework for the assessment of the civilizational importance of the emerging ubiquitous system of computer-mediated communication. Chapter 5 concentrates on the identification of the four central fields of the network's research: "privacy and the public sphere" (for the socio-political dimension of the subject), "identity and the community" (for its social-cultural dimension), "knowledge and the economy" (for its socio-economic dimension), and (more as a cross-sectional topic) "security/insecurity and trust". Chapter 6 considers the theoretical approach of CULTMEDIA while discussing the interdependencies and exchange relations of cultural diversity and new media regarding their importance, consequences and meaning for the process of European integration. Finally, the paper gives an outlook describing ongoing pursuits and wishes for the near future that reflect the present phase of the network's consolidation.

2 Preliminary remark

This contribution describes the foundation process and the conceptual elaboration of a European research network which is dedicated to the socio-cultural consequences and potentials of the new media. It originated from a loose association of lasting bi- and multilateral cooperative working relations and was founded in 2002, at first under the title "new media and culture – the example of the internet" (NEMET). Meanwhile (since 2003) its name has been "cultural diversity and new media – their interaction as an element of European integration" (CULTMEDIA). The members of the network represent a variety of disciplinary perspectives. They come from cultural sciences and humanities, social sciences and psychology, as well as computer sciences and communication/media studies.

Background of the foundation of CULTMEDIA were the results of a study that was carried out by the Office for Technology Assessment (Büro für Technikfolgenabschätzung; TAB) of the German Federal Parliament (Deutscher Bundestag) during the years 2000 and 2001 (cf. Paschen et al. 2002). This study was stimulated by the Committee for Culture and Media of the German Federal Parliament. Its objective was to analyze recent and future effects of the development of new media on the concept of culture, the politics of culture, the economy of culture and the actuation of culture itself. The focus of the study was laid – apart from considerations about changing patterns of cultural understandings and cultural concepts – on the topics "new media and media markets" as well as on "new forms in the production, mediation and reception" in selected cultural areas (literature, music, cinema).

The current changes in the field of the new media and the media markets whose effects were examined exemplarily for traditional cultural fields were pointed out clearly during the study. However, it was also shown that the findings gained in the national frame must be broadened by their European dimensions. In this regard, CULTMEDIA now examines cultural changes in different national contexts while following two priorities: firstly, it refers to the internet as the representative of the new media; secondly, it refers to such cultural practices which change in the life-world and systemic contexts of the everyday use of this socio-technical medium. General interdisciplinary analyses are thereby combined with multinational comparisons (cf. also Banse et al. 2000).

The task of the network is some deeper analysis of the changes of cultural practices (for instance patterns of use, use motivations and use situations) which are connected with the application of the internet. Within the past years, research about network-based communication has developed with a similar pace as the internet itself (cf. e.g., Grunwald et al. 2006). The relevance of this relatively new research field results from the great importance that is attached to the internet by several huge social trends (as for example the globalisation). The internet leads to a new quality of the mediation of humans to humans and humans to nature thanks to its universal character, global nature, high speed and great availability. It takes essentially part in the reconstruction of the world, a fact that is expressed for instance in the emerging tension-relation of reality and virtuality. The alterations resulting from these trends are often assessed as a major cultural change with effects on all fields of life in the modern societies.

Since 2004, the network has been editing a book series of its own, "e-Culture/Network Cultural Diversity and New Media", published by the trafo-Wissenschaftsverlag Dr. Wolfgang Weist, Berlin. So far, 13 volumes have appeared and two more are being prepared.[1]

In order to consolidate itself as a network of cooperative working relations and to strengthen its financial basis, CultMedia is engaged in elaborating proposals for the European Commission, ("Network of Excellence"), the European Science Foundation ("Exploratory Workshop"), the COST-programme (as well as several other national and transnational funding agencies).

3 Overview

In the present chapter, the subject matter of CultMedia is presented, reasons for its research project are provided and further explanations are given. This is carried out in three stages. The first stage (Chapter 3.1) outlines the historical background and provides a general introduction into the relationship of media and culture development. The second stage (Chapter 3.2) explains the research subject of CultMedia in several steps, i.e., the interrelations of cultural diversity and new media and their role in the process of European integration. The third stage (Chapter 3.3) offers a structural overview of the research levels within the network, which also serves as a preliminary outline for further explanations.

3.1 Media & culture – historical background

There has always been a close relationship between the cultural development of Europe's peoples and the utilisation of communication media. Such connec-

1 Cf. http://www.trafoberlin.de/ ⟶ Wissenschaftsverlag ⟶ Reihen ⟶ e-Culture.

tions also apply with regard to the development and utilisation of digital media in connection with that of computer-mediated communication and the transformation of modern societies (cf. Banse, Grunwald and Rader 2002).

The possibilities for societies to elaborate, distribute, store, use and revise information and knowledge constitute an important dynamic factor of sociocultural change. The beginning of script culture, the invention of book printing or the "modern" media such as press, radio and above all TV have made a considerable contribution to cultural and social changes producing serious consequences. According to the information and communication-technological development dynamics, it also makes sense to consider the cultural conditions and implications of the information and communication possibilities represented by the internet, which for the purpose of a better terminology are summarised as "new media". However, it must be considered here that the structural change of cultural practices to be investigated is not merely induced by technology, but is also connected with interrelated social developmental processes; increase in globalisation, individualisation, reflexive modernisation, complexity and contingency as well as a change in values are keywords that occur in this context.

CULTMEDIA now intends to make its research subject cultural changes in the wake of the so-called "new media" with regard to the following two priorities: first, it refers to the internet as *the* representative of the new media, and second, it refers to those cultural practices that are changing in the lifeworld[2]-related and systemic connections of the everyday use of this socio-technical medium. In addition to general interdisciplinary analyses, the main emphasis is placed on multinational comparisons.

3.2 The network's subject of research

Against this background, CULTMEDIA investigates these changes in relation to sociality (formation of individual and collective identities, forms of community building) and culturality (cultural practices and goods as content patterns of social life), which result in connection with the development and utilisation of the internet (as a medium for technical information and communication). Thus the network investigates changes and processes that transform the present European society/societies and that constitute a major part of its transition to a "knowledge-based society".

2 Cf. McIntosh 1997 for an assessment of Jürgen Habermas's distinction between system and life-world (cf. Habermas 1981a, 1981b, cf. also Habermas 1989).

These changes take place in the European Union in an environment of extreme cultural diversity and are thus influenced in their form and process (speed, direction, pattern). Conversely, they simultaneously influence this environment and change the dynamics within which diversified national cultures within the EU develop further in mutual solidarity.

The interactions between "cultural diversity" and "new media" develop in a complex matrix, the salient points of which are set through the terms "culture", "society", "media" and "technology". As far as "society" is concerned, the conditions of influence of social subsystems, the way organisations work and the interaction opportunities of individuals all change in connection with the transformation of their cultural foundations through the extensive use of new media. It is not the sum of the changes in all fields of interaction and life that is decisive but primarily also the changing relationship of penetration and delimitation between "life-world" and "systemic" references of communication and action.

In order to be able to understand the form and importance of cultural transformations which take place in connection with the development of new media, two subjects shall be reflected: on the one hand, the subject of "reality and virtuality" and, on the other hand, that of "space and time". Both subjects are to be considered systematically during the investigation of the three research fields (determined later) and the cross-sectional topic. Here it is decisive that the cultural transformations are not simply a series of extended possibilities of communication and symbol processing or their broader use, but that the new media are essentially characterised by the fact that they make the relationships of virtuality and reality and space and time available and usable to an extent which has been unknown so far and that they "reconstruct" them in a complex way. Instead of supporting one-dimensional theses, such as a "softer" reference to reality through virtuality or a "shrinking" of space through electronic communication, it shall be shown which design potential new media have for creating new relationships, e.g., through the possible access to an enormous reservoir of information sources and stock of knowledge at any time and all over the world.

3.2.1 Homogenisation versus diversification

As far as the interactions of "cultural diversity" and "new media" are concerned with reference to their consequences for the "cultural diversity", it must be assumed that neither the hypothesis of continuous cultural homogenisation nor the hypothesis of continuous cultural diversification can be proved. The concern is rather with an area of tension between the poles of which – homogenisation versus diversification – a terrain of complex overlapping and diverse design options is created. What shape this overlapping takes depends on how it is designed socially, the contextual conditions (language and cultural differences between the European nations) and the forces that act upon it. The form of the interactions between "cultural diversity" and "new media" is not only determined by the characteristics of the "new media", but also by the way in which optional connections become concrete under the influence of protagonist coalitions (network operators, content providers, user groups, media and cultural policy) and macrosocietal processes such as (commercial) globalisation, individualisation or reflexive modernisation.

3.2.2 Fields of research

In order to incorporate the interactions between "cultural diversity" and "new media" in their function as an element of European integration, they are investigated by comparing European nations.

In order to capture the scope and complexity of possible interactions, the implication of cultural transformations is to be explored in the field of social, political and economical dimensions. The three areas of research delimited by CULTMEDIA not only analyse these three dimensions, but also place an emphasis on three research topics that are an important prerequisite for the gaining of an individual understanding of all the current processes of change that are of utmost importance for the further development in Europe.

- Research area 1, "privacy and the public sphere", analyses the socio-political dimension of the topic "cultural diversity and new media" with the help of the following question: What is going to change in the relationship of privacy and the public sphere under the influence of new media?
- Research area 2, "identity and the community", analyses the socio-cultural dimension of the topic "cultural diversity and new media" with the help of the following question: What is going to change in the relationship between identity and community under the influence of new media?
- Research area 3, "knowledge and the economy", analyses the socio-economical dimension of the topic "cultural diversity and new media" with the help of the following question: What is going to change in the relationship between knowledge and economy under the influence of new media?
- In addition, there is the cross-sectional topic "security/insecurity and trust". The research problem of this topic, the changed balance of these two fundamental parameters for the constituency of modern societies,

unfolds in the intersectional plane of the three areas of research. It is investigated by means of the following question: What changes, problems and approaches to solving the problems result with regard to the relationship of security/insecurity and trust in the three areas of research "privacy and the public sphere" (e.g., digital signatures – cf. Langenbach and Ulrich 2002), "identity and the community" (e.g., theft of identities) as well as "knowledge and the economy" (e.g., unauthorised changes)?

3.2.3 European integration

By means of comparative investigation of the interactions of cultural diversity and new media in the three areas of research as well as for the cross-sectional topic, not only the pattern of the interactions in question is to be identified, but it should also be investigated to what extent this pattern varies between European nations. Instead of "only" leading a general discussion on cultural diversity as an element of European integration, for instance the question of whether it is to be understood as a strength or a weakness, the question regarding the connection of cultural diversity and new media is investigated within the context of four concrete complexes which are of vital importance for the development of the present European society/societies and the progress of the integration processes within the EU.

In order to complete this investigation and to contribute to the practical implementation of its results, the CULTMEDIA finally considers what consequences can be derived for well-founded support of the media and cultural policies of the EU and of its member states. The problem here is not only the coherence of its guidelines, but also their approaches and instruments as well as the question of whether there is a special need for action within the framework of the current European expansion, and what measures can be taken to meet this.

3.3 The network's three levels of research

The research work of CULTMEDIA is to be carried out on three levels, which are based on each another but do not necessarily have to be executed in succession, in so far as their interrelation has to be considered. They will be briefly outlined here and described in more detail in Chapters 4, 5 and 6 in connection with the determination of the research tasks related to them, which will be intensively worked on during the later periods of research carried out by the members of CULTMEDIA.

- On the first level, which will be described in more detail in Chapter 4 of this paper, the task is to continue the abstract-conceptional work, to look at it in greater depth and to review it in a reflective manner. This will be achieved by means of state-of-the-art research regarding the interactions of culture and media, which will be presented in the following part together with open questions, and difficulties regarding the posing of the problem. In this context, the fundamental concepts of culture, technology, media and communication[3] will mainly be dealt with. In this chapter they will be explained in two paragraphs that connect the correlations of culture and society on the one hand (cf. 4.1), and those of technology and media (cf. 4.2) on the other.
- On the second level, which will be introduced in Chapter 5, the main problem concerns discursive-comparative research devoted to the current interactions of cultural diversity and new media, especially in the three areas of research introduced above as well as regarding the cross-sectional topic mentioned. First, the area of tension between homogenisation versus diversification will be examined in the context of a discussion of the general connection of media development and cultural change (cf. 5.1). Second, the interactions will be investigated in connection with the three areas of tension and for the cross-sectional topic with the objective being a comparative elaboration of their patterns (cf. 5.2).
- On the third level, which is presented in Chapter 6, the task is to work in a synthetic-operational/conciliar manner, which means developing conclusions, elaborating recommendations and identifying steps for their implementation. Thereby, the role of the interactions (of cultural diversity and new media) as an "element of European integration" is in the focus of attention. In this context, the importance of "cultural identity" (national identity – European identity) for the process of European integration comes first, considered in terms of the connections of their political and economical contextual conditions (cf. 6.1). Second, conclusions will be drawn from the comparison of the above-mentioned interactional patterns between cultural diversity and new media in the three areas of research and for the cross-sectional topic (cf. 6.2). Third, against this

3 Further abstract-conceptional formulations, which are directly connected to the discursive-comparative research level, i.e., regarding the connections of "community" and "identity", "public sphere" and "privacy", "knowledge and economies" as well as "security/insecurity and trust", will be elaborated later. The same applies to the abstract-conceptional formulations for "cultural identity", "European identity" and "European integration", which are directly connected to the synthetic-conciliar level of research.

background recommendations will be developed for the further shaping of European media and cultural policies (cf. 6.3).

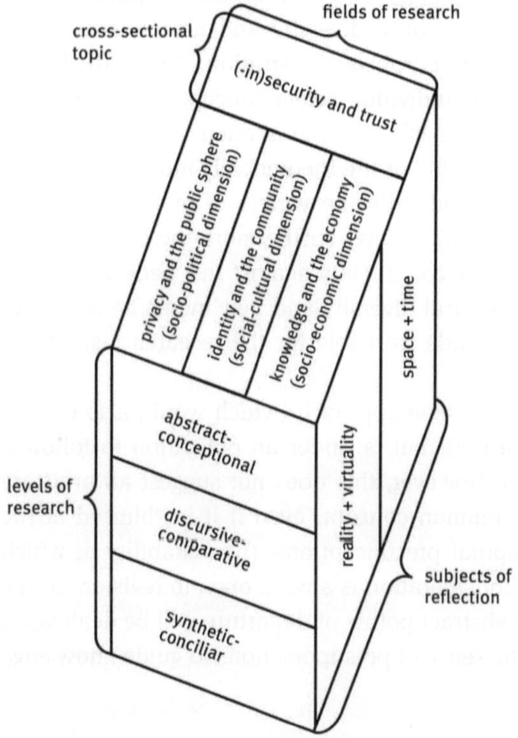

Figure 1: Overview of the scientific design of CULTMEDIA: Research levels, research fields and subjects of reflection (source: by the authors)

4 Abstract-conceptional research-level: Relationship of culture and media

The abstract-conceptional level of CULTMEDIA's research work is the central focus for explanation in this chapter. The interactions between cultural diversity and new media unfold in a complex matrix, the salient points of which are marked by the concepts "culture", "society", "technology" and "media". In this chapter they will be discussed in two paragraphs that combine the connections between culture and society on the one hand (cf. 4.1) and those between technology and media on the other (cf. 4.2).

Concepts are not true or false, but rather more or less appropriate to the purpose for which they are used and constructed. CULTMEDIA's purpose is to work out the interactions between cultural diversity and new media and to comprehend what role they play for European integration. Thus CULTMEDIA cannot be satisfied by identifying an individual, clear concept of culture, technology or media and using this as a basis. As a consequence, the task is not to identify individual concepts as "right" or the theoretically most "advanced". On the contrary, the first goal is to identify the variety of meanings encompassed by these words in science and society. Furthermore the complexity of the discourse should be taken into account. Only against this background can the multifarious interactions of cultural diversity and new media be accessed, consolidated and expanded, i.e., made workable for the research tasks to be dealt with.

CULTMEDIA does not pursue a single approach (which would also contradict its professed interdisciplinarity!), but is under an obligation to follow a variety of scientific perspectives. However, this does not suggest an arbitrary nature: the starting point is a common concept, even if it is "blurred at the edges", including certain conceptual preconceptions, the suitability of which for solving the problems under consideration is always open to revision. In the following section, conceptional-abstract points of departure will be dealt with, which are to be understood in the sense of presuppositions to guide knowledge and actions.

4.1 Culture and society – a changing relationship

As far as "society" is concerned, in connection with the transformation of its cultural foundations, the conditions of activity in societal subsystems, the way in which organisations work and the individuals' possibilities of interaction change through the extended use of the new media. However, it is not the sum of the changes in all fields of interaction and spheres of life that is decisive, but mainly the changing relationship of penetration and delimitation between "life-world" and "systemic" references of communication and action.

"Culture" is conventionally distinguished according to space and area of distribution (regional, national, international cultures), according to social acceptance and/or representativeness (advanced civilisation, subculture), according to specificity of social systems (industrial culture, sport culture), according to the representatives (court culture, cloister culture, youth culture) and according to the bonds with certain types of manifestation (music culture), where the time component is always intrinsic.

For the present it must be assumed that in sum these partial cultures do not fit together to a neat cultural whole, which would make it possible to deduce generally binding orientations. The fact that this cultural variety is reflected in a variety of concepts and understanding of culture corresponds to these results.

With regard to its importance and use, "culture" belongs to the scientifically most delicate and at the same time exhausted notions. In the past decades, the notion of culture has experienced a boom that goes far beyond academic debates. The latest development of the media has taken place against the background of this boom, and what is regarded as possible consequences of media development on the understanding of "culture" depends on the historically developed concept of culture that is used as a basis.

In this context, first of all the most important types of notions of culture have to be distinguished. A commonly used differentiation is made between ethnographic or ethnological and those strongly normative notions of culture from the tradition of humanism (cf. King 1993: 2). Another possible differentiation is made between "culture" as a way of living in general (ethnological concept of culture), as art, as a humanely organised way of living (committed normative concept of culture of humanistic tradition) and as a symbolically negotiated sphere of values and norms in society (cf. Fuchs 1999: 220–221).

According to Siegfried J. Schmidt, the present boom of the notion of culture in science and politics is not simply a current fashion, but rather an "indication of a significant social development", a "development from the dominance of material factors to a dominance of knowledge", which, in turn, is considerably influenced by the development of information and communication technologies (cf. Schmidt 2000: 32–33).

4.1.1 Culture – territoriality – globalisation

A striking characteristic of new media is the fact that their spread changes the cultural importance of spatial closeness and distance. The widespread view is that the networked individual exceeds with its actions in the communication network beyond the borders of nationally constituted societies, and the possibilities of transnational cultural exchange and individualisation increase. This is promoted by the general tendencies of globalisation, which give a great deal of significance to the current change of concepts of culture.

Due to various stimuli, in the past years new approaches to investigating the relationship between "western" and other world regions have developed and so, too, has an increased interest in individual and collective acquisition processes of cultural contents distributed by mass media. The importance of these acquisition processes for the cultural change in the different world regions may lie in the alternative (detached from the traditional spatial context of experience) life plans which become available, in the "possible lives" which are conveyed by the media (cf. Appadurai 1996). The possibilities of computer-assisted communication of those who are spatially far apart and the increasing cultural importance of migrations will probably result in a further increase in the importance of ethnological topics – such as "virtual ethnicity" and "cultural hybridisation".

In the debate about traditional notions of culture, new concepts of culture are sometimes created that – although they are often not thoroughly outlined and only elaborated as first attempts – show some common characteristics: the individual is placed in the centre of the cultural-theoretical interest again, where, among other things, special relevance is attributed to post-traditional forms of community building and transnational links. There is a tendency of "liquefaction" and "de-territorialisation" of the notion of culture, and special emphasis is placed on aspects of the connection between cultural change and cultural globalisation. If this tendency is continued, a central topic of cultural-political debates could be the question of what role a future "transnational state" (cf. Beck 1997) could play regarding the global "cultural space of the internet".

Processes of individualisation as well as globalisation cause a relativisation of the cultural relevance of the social relationships that are characterised by spatial closeness. We are no longer dependent on the local community as a source of information, experience, entertainment, a feeling of security and the way in which we see ourselves to the same extent as before. According to widespread belief, the tendency of relativisation of cultural relevance of spatial nearness is closely connected with the latest development of the media, but also with migration and tourism.[4]

Thus polarisations such as "globalisation versus localisation", "homogenisation versus heterogenisation" or "universalism versus particularism" do not lead us anywhere. The fact that they still characterise the debates on the interactions between globalisation and media development is probably due to the novelty of the developments in question.

4 However, the perception of this tendency should not tempt us to neglect local aspects. There are attempts to take this into account with concepts such as that of "glocalisation" (Robertson 1998).

4.1.2 Towards a modern notion of culture suitable for the project

The answer to the question of what is to be understood by an up-to-date notion of culture therefore becomes a central starting point of CultMedia. In order to back up CultMedia's work on the discursive-comparative (second) research level, a notion of culture is needed which allows an understanding of cultural diversity in such a way that it is not only possible to pick out the interactions between cultural diversity and new media as a central topic, but also especially in the four areas of research chosen.

It is natural to choose a relatively broad notion of culture in this context, such as the one from the final declaration of the World Conference on Cultural Policy held in Mexico in 1982.[5] However, its notion of culture is too general, not specific enough, and above all, cannot be operationalised. Only a sufficiently broad notion of culture opens up the possibility of observing how the media changes influence cultural structures, organisational forms and ways of communication and interaction ("communicative action", truth, authenticity, etc.).

Despite this, the starting point of our further considerations is less extensional and more genetic: culture is formed within processes of cognitive, emotional and practical debate between interacting individuals regarding the conditions for their actions, which, on their part, not only include social, but also technical and ecological dimensions.

The following definition, which is to be understood in the sense of a first approach, could be appropriate for the concerns of the CultMedia project: according to this definition, culture is the result of the human being's coping with life and existence in an action-based and communication-based community, in other words, the "spatially-temporally restrictable totality of common material and non-material products, internalised values and interpretations of meanings as well as institutionalised forms of human life" (Klein 2000). Thus if culture is understood as "the totality of conscious and unconscious collective patterns of thinking, feeling and acting, which are socially acquired and handed down by people as members of a society and constitute a specific, characteristic of this society that can be delimited" (Hermeking 2001: 18), then this follows both the summarising and "classical" considerations of Alfred L. Kroeber and Clyde Kluckholm, who explicitly include "the achieve-

5 There it is stressed that "the Conference agrees: that in its widest sense, culture may now be said to be the whole complex of distinctive spiritual, material, intellectual and emotional features that characterize a society or social group. It includes not only the arts and letters, but also modes of life, the fundamental rights of the human being, value systems, traditions and beliefs" (UNESCO 1982).

ments expressed in the material goods produced" in culture (cf. Kroeber and Kluckholm 1952: 181), and Arnold Gehlen's starting point for whom culture is the "epitome of materials/equipment, of expert technology and thinking techniques" (Gehlen 1958: 86).

Determining parts of this understanding of culture are mainly:

- non-material as well as material (concretised, "reified") areas;
- handed-down, long-lasting productions;
- both spatially ("community") and temporally limited or delimited (also: distinguished) material and non-material "patterns";
- the aspects of being both, a product resulting from actions and an element that takes part in the formation of further actions.

4.1.3. Cultures of privacy and the public sphere, of identity and community, of knowledge and economy, of security/insecurity and trust

Particularly in this sense, culture is more than the sum of products of an unclearly differentiated societal area of action, which has specialised in the production of cultural goods, in which "creative artists" work, in arts, music and literature as well as in further branches of this profession. But in this particular sense culture represents more than simply lifestyles or a set of norms, values or convictions. It is rather to be understood as a kind of matrix that reveals areas of meaning, that offers certain possibilities of association (and excludes others), that provides meaningful explanations for distinct patterns of behaviour and interaction, that works by linking together descriptive and prescriptive aspects, thus indicating cognitively correct ("functional") and normatively correct ("good") behaviour, or better, opportunities of behaviour.

This cultural shaping can also be observed in CULTMEDIA's four areas of research:

I. Which distinct communicating and acting is, can be, may be and should be executed "in private" – i.e., not in public – and which distinct communicating and acting even has a right to be shielded and protected, and where the limits of this shielding and protection are, is culturally coded. Conversely, it is culturally defined and explained which acting and communicating should, may or can be executed "in public", which "public" conduct deserves protection and support (e.g., freedom of assembly) and which does not (e.g., revolt and breach of peace).

II. dentity", i.e., on the one hand the possibilities of individual identity formation of subjects of action and on the other hand, their consideration or appreciation on the part of social communities or society as a whole, also follows this pattern of cultural coding. As social structures, sexes, professions, background or nationality represent a kind of matrix for the development and formation of distinct ideas and ways of behaving which show a certain degree of stability, but which can also be formed, changed, opened up for new things or shielded against others. The same applies to religious or political convictions (republican, monarchist) or "cultural" forms of expression (e.g., punk, skin) and areas of interest (hacker), identification with a sport (e.g., football fan), with certain teams (e.g., Real Madrid) or athletes, with certain music styles (e.g., reggae), bands, stars or forms of life (e.g., "inhabitants" of a digital town in a "Multi User Dungeon"). In this way we express what we believe in (contents), how we act (conduct) and how we express this (clothes, hair styles, flags, signs of identification), but also the character (aggressive, peaceful, showing solidarity, stressing competition) of the "community" which the individuals with their corresponding subjective identities belong to.

III. As a consequence, culture also influences the "knowledge" of a society, i.e., the forms of knowledge as well as their acceptance as legitimate forms and sources of knowledge, the contents of knowledge as well as their acceptance as valid, correct, true, and ultimately the ways in which it is used. "Acting in economic contexts and with economic goals" is also not at all a culture-free conduct determined by its function, since culture defines its purpose, determines what is of value and benefit, conveys its forms, its patterns (cooperative, competitive, etc.) and what are transferable commodities (goods) and what are not (love, human dignity, public goods).

IV. What is considered as an insecure or secure state is not fixed, but depends on cultural preconceptions. In this way situations and contexts of action are assigned attributes of "insecurity" or "security" to a varying degree, which leads to the consequence that one behaves differently (carefully) in these situations or contexts of action. In connection with this, "trust" but also "distrust" represents a necessary resource for any behaviour (intending success). In turn, both are influenced by culture.

4.2 Technology, media and computer-mediated communication

For the understanding and the notion of technology – analogous to the notion of culture – it is of utmost importance to be positioned broadly and with enough complexity to be able to work on the embedding of technology in cultural processes and the interweaving of technology and culture systematically, on the one hand, with the effects of technical change on culture and, on the other hand, with the influences of culture on technology. The conditions under which media develop and, in turn, become effective are inextricably interconnected with these contexts.

4.2.1 Cultural contexts: Technological artefacts and technological action

Common "definitions" of technology are for instance: "[...] technology describes artificial things and procedures which serve practical purposes" (Sachsse 1992: 359). Such formulations – which should be categorised as "understanding of technology in the narrow sense" – see the concrete, "artefactual" aspect of technology as the central focus. This is rather one-sided, since the question of the origin of technology is not touched on. Technology has not been "given" to man (like nature), it is not – in the original sense of the meaning – "natural" and "does not fall out of the blue", but it must be "made", "produced", "created".

Only against this background does it become obvious that technology is not "natural", but "artificial". In addition, it must be said that specialized technical systems represent the means for the realisation of human purposes. For an appropriate understanding of technology both aspects have to be considered. In such an expanded understanding of technology (notion of technology of a "medium" range) technology includes, first, the quantity of user-oriented, artificial, concrete objects (i.e., the artifacts or technical expert systems), second, the quantity of human action and facilities in which expert systems are created, and third, the quantity of human action in which expert systems are used (cf. Ropohl 1993: 672). If understood in this way, "technology" not only describes the objects made by man ("artifacts"), but also includes their creation and utilisation connections ("contexts") (i.e., the aspects of "having been made", "been applied" or "been used"). Thus technology is not considered to be something static but belongs to an area with genesis, dynamism and change (cf. Banse 1999).

If we now consider that in the contexts mentioned, different conditions (especially of individual, scientific-technological, economical, legal, political, ecological and ethical nature) are of influential importance, it first becomes obvious that by means of this "broad(er) understanding", technology, if understood not as an isolated, autonomous area that seems to constitute an objective reality, becomes a "social" phenomenon in terms of its development which is insolubly linked ("networked") with the individual and society, with politics and economy. Second, it must be understood that technology "finds its application and everyday use [...] in a socio-cultural context, in a context of collective interpretations and explanations" (Hörning 1985: 199). The starting point is the understanding that technical objects do not necessarily have to be the way they are, and the way they are omnipresent to us, i.e., how they reach our everyday life from autonomous technical conditions. In their creation as well as in their technical use, expert systems are an expression of both, own intentions and purposes and those of others. Despite all exactly installed and registered instructions for action, which promise the optimum utilisation of the function especially for the layman, particularly everyday technology often offers tremendous freedom of utilisation, too: taken up by one, badly used by another, ignored by a third person – but always against the background of certain utilisation expectations, influenced by evaluation and advertising as well as embedded in certain societal and technical "infrastructures". The "usefulness of technology is always also something interpreted culturally" (Hörning 1985: 200). Thus it becomes obvious that culture considerably influences the implementation and diffusion of technical solutions via the people "carrying" it, by e.g., using or not using (refusing) it for the realisation of purposes, by forcing modifications, improvements, adaptations and by creating "regulations" for the behaviour of man-technology-interactions. Therefore, in CULTMEDIA questions should be asked regarding the degrees of cultural freedom in the reception and during the handling of technology in everyday life, and also regarding the fact how different groups and generations with different ways of life and cultural orientations handle offers of technology. These offers of technology maybe but do not necessarily have to be identical.

4.2.2 Media: Their technological and their socio-cultural side

One difference in orientation running through the media discourse is to understand media rather as technical systems on the one hand, or as socio-cultural practices on the other hand. However, for CULTMEDIA's intention it is important to consider both the technical and the socio-cultural side of the media since in their interactions lies one of CULTMEDIA's relevant reference problems. In order to develop an adequate understanding of media for CULTMEDIA in this sense, it is important to search the various technical media concepts and to clear up conceptional one-sidedness.

For the main question regarding the relationship of new media and culture – as generally in discourse on technology and society – the question arises as to what extent societal or cultural changes are necessarily anchored in technology itself ("required" by technology as so-called adaptation services) or, vice versa, to what extent changes are dependent on social and cultural practices into which technology is integrated as a means. While processing these questions we have to be aware of the tendency that there exists the danger of oversimplifying our considerations following the one or the other of two – too one-sided – conceptualisations. Regarding the (new) media a (rather) media-theoretical and a (rather) communication-theoretical one-sided conceptualisation can be distinguished (cf. Rammert 2000).[6]

Both conceptualisations are illustrated in media reality. For CULTMEDIA's project, a decision in favour of one of the perspectives is neither useful nor necessary. They can be emphasized as two different and from case to case heuristically useful conceptions of the dynamism of socio-cultural change inherent in the relationship of media and culture.[7]

Medium means central part, mediating element. In communication science it describes – deduced from this – an agency for mediating information. Thus media are mediation systems for information of all kinds (news, opinions, entertainment). Their function is the transport of contents during which specific restrictions of the medium may affect the form of the content. Technical means of communication (e.g., in the form of information and communication technology) serve for the transmission, storage and distribution of information.[8]

6 The media-technical concept is based on the line of thought that the medium itself changes the relationship of man to the world independent of its use. The exaggeration of this line of thought is: every medium creates another view of the world. In this exaggeration lies a highly simplified understanding regarding the connection of media and communication, namely that every new media technology leads to another way of communication and supersedes the old forms of communication. The communication-theoretical concept results from the exaggeration of the line of thought that the medium is a neutral means which opens up new possibilities and that it is only up to the human actors and their actions how communication and interaction relationship change through the use of the media. This line of thought reduces the reality of the media to the handling of tools.

7 One-sidedness can be avoided when analytically subsequent levels of dialectics of media structures and media practices, which were proposed by Werner Rammert, are considered equal and in their interactions (cf. Rammert 2000: 125; transl.): (a) the *media* "as material carriers with specific reference to the human senses and to the physical environment, to temporal and spatial dimensions"; (b) the technical forms, "as they are concretely constructed in technical projects, configured in their material, physical and symbolic elements and installed in their context"; (c) the *institutionalised forms* "how they are legally and technically standardised regarding their functioning and their handling"; (d) the *programmes*, "the functions and services offered"; (e) the *practices*, "the way in which manufacturers, providers, operators and above all users handle the media".

8 If one searches for a common denominator in media-scientific literature, it can be found in the fact that media serve for the storage and representation of information and the mediation of communication. Communication is usually understood as an exchange of information between individuals, which is based on mutuality and interaction, whereas

An important differentiation in the connection of media and technology is that between primary media for the functioning of which the use of technology is not necessary (e.g., theatre), secondary media for the functioning of which the use of technology is necessary as far as production is concerned, but not for reception (e.g., daily newspaper), tertiary media for the functioning of which the use of technology is necessary for both sides, i.e., for both production and reception (e.g., record). In addition, there are quaternary media for the functioning of which besides technical support of production and reception, the technical mediation of the distribution is indispensable (online media, which are suitable for dissolving the conventional receiver/sender relationship) (cf. Faulstich 2000: 21).

As a working definition, media are described in this context as those socio-technical systems and cultural practices of the dissemination and storage of information which serve for the design of communication and interaction and thus influence the collective as well as individual perception and the building of experiences in the real world. As another temporary classification, new media describe those media whose technical foundations are based on digitalisation, miniaturisation, data compression, networking and convergence.

4.2.3 Digital media: Their convergence and the genesis of a ubiquitous communication system

With regard to the understanding of media as "channels", about twenty individual media can be differentiated today (following Faulstich 2000: 22), namely – in alphabetical order – books, computer, internet/DVD, journals, letters, magazines/pamphlets, newspapers, sheets of paper, sound radio and television.

The "old" media were once "new", too. Therefore, the category new media should be characterised substantially beyond the aspect of their general everyday use which is still to come. Furthermore, a suitable notion is to be included in this substantial characterisation, which clarifies what is meant by new media in CULTMEDIA's narrow research context.[9] Without a current satisfactory

---

information is understood as a process in only one direction. Neither informing nor communicating is necessarily modelled as simple (linear) "sending and receiving". Messages should not only be considered as physical events, but must also be understood as "coding" and "decoding", i.e., as interpretative acts in a social "environment" and against a cultural "background".

9 Regarding the aspect of their material carriers or their object area the new media can be roughly delimited as an area of microelectronically based hardware and software technologies. "The heart of these technologies are highly efficient computers of various forms the peculiarity of which is, among other things, that they are not limited to working in isolation, but that they can be linked almost without restriction both locally [...] and regionally and

solution for this task, the fundamental importance of convergence should be stressed in this matter.

Against the background of this medial convergence, CULTMEDIA is focusing its investigations on the relationship of cultural diversity and new media on the internet. Currently the internet is not only the dominating and most dynamic element in the area of new media, but it is the element with quantitatively and qualitatively the most far-reaching consequences of socio-political, social-cultural and socio-economical nature.

Internet means "interconnected networks". It is a combination of computer sub-networks in which digital data is exchanged in the form of packages (TCP/IP). The internet connects more or less autonomous computers for the purpose of exchanging data. The universality of the computers characterises the network and gains a new spatial-temporal dimension with the global networking. The internet is the synergy of computer and global network integration: a hybrid, partly computer network, partly network computer, partly individualising the user intimately, partly socialising him.

The internet gains its efficiency and its ability to develop from:

- the universality of the computers that constitute it;
- the ability to digitalise all information and treat it as data;
- the efficiency of data exchange in the form of packages;
- the setting of standards that work on all computers;
- the easy handling of the web browsers as a universal computer interface;
- the efficiency of the hypertext technology which is able to embed complex data structures unobtrusively;
- the interactivity of the WWW and the possibility for the user to have a creative input on it;
- the high level of distribution.

Even today the internet modifies, supplements and/or replaces a variety of historically developed cultural techniques: it is a(n)

- individual means of communication (telephone, videophone, email, chat);
- medium of entertainment (film, television, radio, online games, etc.);
- medium of information (newspapers, books, magazines, libraries, databases);

ultimately worldwide, and, at the same time, allow equal mutual exchange ('interactivity'). Consequently these devices become an aid for individual and transindividual communication in all spheres of society" (Boehnke et al. 1999: 9; transl.).

- medium of knowledge processing which – supported by language recognition and synthesis, recognition of (hand)writing and classification of graphic data – leads to a multiplication of intelligence services;
- medium of trade (buying, selling, auction);
- medium of direct distribution of non-material goods (print media, music, films, financial transactions, education, consulting);
- medium of remote control, remote maintenance, remote supervision, i.e., an “integrated computer”;
- medium of self-presentation and self-fulfilment;
- medium which modifies the users’ spatial-temporal coordinates through its global distribution and its fast data flow.

Formed as an integrated network, the internet is able to integrate all traditional networks (within the framework of their digitalisation) and all newly created networks (UMTS, etc.). Thus the internet qualifies itself as an outstanding means during the analysis of cultural changes which are connected with the emergence of new media. Another reason for CULTMEDIA’s focus on the internet in the investigation of the relationship between cultural diversity and new media lies in the importance that is assigned to the internet in public debates, as an engine of societal and as a motivating force of cultural change.

4.2.4 Subjects for reflection, characteristics and significance of new media

In order to be able to better understand the form and importance of cultural transformations which result in connection with the development of the new media, two subjects of reflection should be used, on the one hand, the relationship of “reality and virtuality” and on the other hand that of “space and time”. Both subjects of reflection are to be considered systematically, i.e., on all three levels of research of CULTMEDIA as well as its four areas of research. The decisive factor here is that the cultural transformations are not only a consequence of the extended possibilities of communication and symbol processing or of their broader use, but that the new media are mainly characterised by the fact that they change the relationships of virtuality and reality and of space and time and make them available and designable, “rework” them in a complex way to an extent that has been unknown so far. Instead of supporting one-dimensional theses, e.g., that of a “softer” relation to reality through virtuality or a “shrinking” of space through electronic communication, the topic should be what design potential is inherent in the new media to create new conditions, e.g., through the possible access to an enormous reservoir of sources of information and stocks of knowledge at any time and all over the world.

The novelty created by the digital media can be seen in the various possibilities to realise different communication patterns equally well by means of this one medium of convergent new media: 1:1 – one person can communicate with another; 1:n – one can turn to many; m:1 – many people turn to one individual; m:n – many people turn to many others. On this basis, new mixed forms are possible, e.g., if in an email to an individual addressee through "CC" a number of people can also read it and thus participate in the sender's community for a particular topic. In this context, the notion of "virtual community" is also of importance, but here the core and the periphery, that means the "brief visitors" of such communities, must be differentiated.

Digitalisation makes it possible to extensively combine data in a new way, because it means the coding of information in discrete sign strings. These, and with them the digitalised information, can then be stored and communicated without any loss, but they can also be taken to pieces and combined and transformed in a different way. In addition to this, it becomes possible to manipulate them almost without any traces, to store them in a practically uncontrollable manner in unknown places and to open them in unnoticed manner. Digital information technology largely detaches the discrete sign strings from the material carrier by storing and transmitting by means of binary encoding. Almost absolute flexibility while handling digitalised information becomes possible. The other side is the emergence of new forms of dependence on technology, e.g., on the latest programme version or technology generation.

The internet, as the basis of online information and communication is – as a global transmission system which changes media development comprehensively – of special interest for CultMedia. Within the media and communication sciences the determination of online media is still disputed. A differentiation between mass, group and individual communication starting with the relationship of production and reception has to be ruled out, because online media detach the "role asymmetry" of sender and receiver. For an investigation, various forms of communication in the internet and the internet services have to be differentiated.

In this context, changes of identities, social relationships and social groups through the internet are of special relevance. In the following part this is to be demonstrated as an example regarding the statement that via internet an interpersonal exchange can take place which is considerably different from personal meetings. For this, the term, "computer-mediated communication" (CMC) has been established, which so far is employed to mean communication conveyed by text. For differentiating the different aspects of CMC, Nicola Döring refers to the natural basic form of human communication, the face-to-face situation: in this situation we are in the same place at the same time (physical co-presence) and exchange verbal and non-verbal messages. Communication media free the interpersonal communication of the restriction (or the pleasure) of co-presence and make it possible to establish contact with people in spatially different places (cf. Döring 1998: 34).

The differentiation between two forms of communication is widespread: time-delayed, asynchronous communication (e.g., letter, email, fax, telegram, message on the answering machine) and simultaneous, synchronous communication (e.g., phone calls, chatting, video conferences). With CMC – especially through emails, newsgroups and WWW sites – asynchronous and – particularly through "Internet Relay Chat" (IRC) – synchronous connections are possible. Of special significance and often discussed is the possibility of anonymous communication in the internet.

With synchronous CMC, the communication processes change fundamentally because a virtual level of action comes to fruition. People not only communicate but also interact virtually, e.g., in the game world of the "Multi User Dungeons" (MUDs).

People who do not know each other in the classic sense communicate with each other across national and cultural borders. The anonymous form of communication which is possible through the internet detaches the forms of civility in the communication among the people present. In the chat rooms or newsgroups in the internet, which determine most images of electronic and virtual communities, various groups come together via the topic. Information between a large number of senders and receivers is transported there. The groups react to the lack of civility, which is caused by the anonymity of communication, with their own common rules of conduct (e.g., netiquette or online slang). In the case of business transactions via internet, anonymity and the (non-)binding nature present a problem to which people have reacted by creating technical substitutes for conventional forms of business civility (e.g., electronic "signatures").

The hypothesis that new possibilities of cultural participation and political organisation result from media development is of special importance to CULT-MEDIA. However, the main emphasis here is often placed on people in the poorer countries of the world and less on the lower strata in the rich countries. Here, an investigation of the internet conducted by James Slevin provides an interesting example, which is, among others, based on theories by Anthony Giddens (cf. Slevin 2000).[10] This study is particularly interested in the ques-

10 Giddens sees a close connection between the processes of the increase of spatial-temporal distance, the disembedding of the social systems (the "lifting" of social relationships from fixed interaction connections and their unlimited restructuring across

tion of how the internet changes the relationship between local activities and the "(inter)action across distance" and how it can be used to better control the influence of globalisation processes on society/societies. Slevin understands globalisation as an "inherent feature of the modern world" the origins of which reach far back from an historical point of view. Globalisation currently causes a new formation of space and time during which action is of central importance across distances. In this context the internet plays an extremely important role because it can make this action much easier.

New media change the cultural importance of spatial nearness and distance. The linked individual grows – according to a widespread opinion – with its interactive and communicative action beyond the borders of the local community and national society and can participate in transnational cultural exchange. The current media development mainly changes the culturally important functions of spatial (but also historical-temporal) nearness and distance, such as in the case of the decoupling of spatial nearness on the one hand, and the formation of cultural communal spirit as well as the assertion of cultural identity on the other hand (keyword "de-territorialisation of culture"): on the basis of the mass media, territorially localised, and at the same time compact cultural traditions can become insularly distributed. Thus the important role which the mass media has played in the social construction of "imagined communities" (including the modern nation-state) for three centuries is changing due to the current impetus for globalisation in the field of the media. Moreover, new media are becoming helpful now for the identity policies of small and specialised kinds of communities and of other culturally relevant social clusters (e.g., new social movements, non-governmental organisations, transnational concerns and ethnic, sexual and other minorities).

5 Discursive-comparative research-level: Interactions between cultural diversity and new media

In this chapter, CULTMEDIA's discursive-comparative research level is dealt with, which comprises further research subjects regarding the topic of "Interactions between cultural diversity and new media". They will be introduced, first,

space-time-spans) and the globalisation which is characteristic for the modern age. According to his view, the three processes mentioned were and are basically influenced by the development of the media (cf. Giddens 1990).

in connection with a discussion of the field of tension of cultural homogenisation and diversification (cf. 5.1). They will be further explained in connection with the identification of three areas of research and the cross-sectional topic; in this respect the comparative analysis of the interaction pattern of cultural diversity and new media will be focused on the complexes given (cf. 5.2).

Before the interactions between "cultural diversity" and "new media" in their function as an element of European integration can be tackled critically (cf. 6), they must first be analysed, described and structured. Through this process, their dynamism and their patterns will become critically incorporated.

Qualitative analyses of both the basic cultural conditions and implications as well as the cultural consequences and effects of the new information and communication possibilities, a synonym for the internet, should be the main emphasis of CultMedia's work. The question of the "culture of the internet", which can be interpreted both in the direction of the basic cultural conditions and processes which have characterised the internet and in the direction of the new cultural forms of network-based communication, which are especially characterised by the internet, represents a continuous interest in knowledge.

5.1 Homogenization versus diversification

With regard to the interactions of "cultural diversity" and "new media", and more precisely, their consequences for "cultural diversity"[11], it must be assumed that neither the hypothesis of continuous cultural homogenisation nor that of continuous cultural diversification can be proved. In this respect it is rather an area of tension between the poles of which – homogenisation versus diversification – a terrain of complex overlapping and diverse design options is created. The shape of this overlapping depends on how they are designed socially, what contextual conditions (language and cultural differences between the European nations) and what forces act upon it. The form of the interactions of "cultural diversity" and "new media" is not simply determined by the characteristics of the "new media", but also by the way in which optional connections become concrete, under the influence of protagonist coalitions (network operators, content providers, user groups, media and cultural policy) and macrosocietal processes such as (commercial) globalisation, individualisation or reflexive modernisation.

11 The reverse direction of the topic, which regards the interactions of cultural diversity and new media as far as their consequences for these media are concerned ("Are they implemented, used, developed differently?"), will only be dealt with in second place.

In the context of the interactions of cultural diversity and new media and their role in the process of European integration it is of special interest to see if and to what extent there are similarities and patterns in the relationship between Eastern and Western Europe or between Northern and Southern Europe, whether similarities and differences form a pattern or if they are distributed in a particular arrangement.

5.1.1 Intercultural communication/dialogue and cultural heritage/memory

For the purpose of a further analysis and the preparation of subsequent considerations on the synthesis, it is useful to differentiate between intercultural "communication" on the one hand and intercultural "dialogue" on the other hand. "Intercultural communication" describes communication that takes place between actors who belong to different cultures. "Intercultural dialogue", on the other hand, is understood as a special form of *meta*communication, namely communication in which the focus is on "intercultural communication". Intercultural dialogue is characterised by the fact that intercultural communication becomes reflexive, i.e., by the fact that "intercultural communication" becomes an explicit topic in processes of intercultural communication.

Against this background of differentiation the problem that an increase of processes of intercultural communication may lead to increased understanding between the cultures, although this is not necessarily the case, can be looked at. When intercultural communication takes place, it is certainly a prerequisite for the development of mutual understanding and mutual acceptance. However, the fact that intercultural communication takes place is no sufficient reason for the fact that it also happens in this way. It is much more likely to be the case that non-understanding and non-acceptance (nevertheless) prevail or (even) will prevail.

Originally, cultures are related to space or territory because communication processes which build up cultures first only appeal to presence, or more precisely, the accompaniment of interaction processes between co-present actors. Even under these conditions, intercultural communication is possible, but – due to the condition of co-presence – it presupposes the mobility of the actors, meaning their physical movement through space and time. Intercultural communication – as communication between actors of different cultures – becomes (increasingly) easier and less costly if transmitted via media (letters, books, radio, CMC), because these media can bridge space and time (as distances which separate cultures) without having to transport the actors. Instead, statements or, in other words, "messages" are transported (across cultural borders). The more communication takes place in this way across space and time, i.e., bridging the condition of space/time co-presence, the more "intercultural communication" can take place. At the same time, the (media-specific) conditions of the possibility of understanding the contents of intercultural communication processes change with that. Due to the lack of co-presence, common environmental conditions that make interpretation processes easier do not apply. In addition – due to the lack of co-presence – one cannot react directly or ask how something was meant if uncertainties occur regarding the intentions for the action or what was communicatively meant. However, under the conditions of the medium "cyberspace", some of the restrictions of other media do not apply. In the internet people can communicate both with a time delay and simultaneously. With CMC it is possible to create a virtual space of communication that simulates or virtually facilitates space/time co-presence.

What status does "intercultural communication" have in connection with the interactions between "cultural diversity" and "new media"? What course does "intercultural communication" take and what consequences does it have as an exchange process between cultures of the relationship of privacy and the public, cultures of the relationship of identity (formation) and community (building) and cultures of the relationship of knowledge and economy/economies as well as cultures of the relationship of (in)security and trust?

What status does the topic of cultural memory, cultural heritage in connection with the interactions between "cultural diversity" and "new media" have? Which role do the digital media play for its conservation, for the free access to it, for its use in innovative cultural practices?

"Cultural heritage" and "cultural memory" play an important role for all three subject areas, in the socio-economic area as a non-material source of prosperity and Europe's economical ability to develop ("Knowledge and the Economy"), in the socio-political area regarding the ideals of order and freedom ("Privacy and the Public"), in the socio-cultural area ("Identity and the Community") regarding the hearts of identity formation processes (subjectivity, sense of community, common metaphysical convictions, patterns of interpretation).

5.2 Interactions of the three research fields including the cross-sectional topic

Interactions between the (respective) cultures ("cultural diversity") and the "new media" (internet, digital media, multimedia, convergence) take place in all areas of society and in all European countries. In order to be able to identify and analyse them, these interactions are to be investigated intensively and comparatively in three areas and a fourth cross-sectional topic.

5.2.1 Research field 1: Privacy and the public sphere

The first research field of CultMedia is called "privacy and the public sphere". In terms of the relationship between cultural diversity and new media the topic here are the conditions of the relationship of privacy and the public sphere, and whether they change through computer-mediated communication. "Privacy" comprises the right of every individual to protect their private sphere. This basic right is a part of the constitutions of modern societies and is expressly recognised in various international agreements. Since the beginning of modern times it has functioned in the sense of a regulatory idea as a basis of social life, interpersonal communication and social interaction. This right results from the interest of the individual in maintaining a space which is supposed to be free of unauthorised inspection and influence through other people, companies, institutions or state organs. A dimension of "privacy" regards the personal conduct and preferences of action in the sense of sexual, religious or cultural practices. In addition, there is the dimension of personal communication conduct when using the respective media and services (telephone, email, chat, etc.)

Finally, there is the dimension of accessing personal data (data protection), for example of a medical and financial nature, which are stored, transmitted and processed digitally. The "public sphere" describes that societal area which, in contrast to the private sphere (the individuals, the family, economic decisions based on private property), should principally be open to all people belonging to a society or which must be open to all members of society as a necessary prerequisite of democracy. The border between private and public sphere is variable. It is subject to historical changes, determined differently in different cultures (e.g., through everyday practice or public opinion) and is fixed through political-legal decrees in one way or another. In addition, it is dependent on the technologies and media that are available and used.

The research field can be outlined through the following points:

- changes in the perception of privacy and the public sphere;
- influence of new media (especially the internet) on the shifting of the understanding of the borders between privacy and the public sphere;
- influence of the new media on changes in the value system of the individual regarding the relationship of privacy and the public sphere – universalism and particularism;
- possibilities and challenges of digital media regarding the authenticity and trustworthiness of pictures and documents or their sources in connection with their cultural meaning and their influence on the formation of (political) opinions and intentions.

5.2.2 Research field 2: Identity and the community

CULTMEDIA's second research field is called "identity and the community". In the context of cultural diversity and new media, the problem here is the changing pattern of individual identity formation (in the relationship of personal and social identity) as well as the formation of communities (as an association of people who live together, interact or communicate with each other). Of further interest is the release of the spatial-temporal neighbourhood, to, as a consequence, changed norms of conduct as well as to changed forms of inclusion and exclusion and mainly to the question whether they lead to any consequences for the social differentiation and cultural diversity.

The research field can be outlined by the following aspects:

- effects of development and introduction of technologies of computer-mediated communication (CMC) (internet, WWW), digital media (DM) (digital photo- and videography) and those for the creation of virtual realities (VR) on communication, interaction and perception processes, cultural practices and the way we see ourselves and reality;
- the question how the use of these media and technologies changes the currently practised (experienced, seen, created) relationship of virtuality and reality (perception, representation, construction, simulation, deception, space/time subject) and our (everyday, lifeworld/professional, scientific-technological/and philosophical, reflexive) understanding of these conditions;
- problem of determining the significant – differentiating – characteristics of "computer-mediated communication" (CMC), digital media (DM) and "virtual reality" (VR) technologies within the framework of their application and use, especially in comparison with the use of analogous technologies (fax, telephone, telegraphy) as well as mass-medial communication (radio, cinema, television), of written communication (letters, books, newspapers) and the real-world "face-to-face" communication (spoken, written, visual and hypermedial communication / uni-, bi- and multimedial communication);
- changing patterns of individual identity formation (ano- and pseudonymisation; presentation of changed physical, sexual, personal, ethnic and social characteristics) through the use of the internet and the media (cyberidentity) and their consequences for interpersonal and social relationships (delimitation or continuation in "real life" (RL); opening of communication or narcissistic and autistic tendencies, etc.);
- changing patterns of collective identity formation (digital citizenship) and building of new communities (virtual communities) through the use of the internet and the media (centralisation of interests; affectual compo-

nents; untying of the spatial-temporal neighbourhood; norms of conduct; inclusion and exclusion) as well as their consequences for the social differentiation and the cultural diversity in general (sub-cultural, ethnic, regional and national identities; varieties in languages).

5.2.3 Research field 3: Knowledge and the economy

CULTMEDIA's third research field is called "knowledge and the economy". In the context of cultural diversity and new media, the problem addressed here is the changing interaction of conduct which creates (scientific) and uses (economic) knowledge. To the extent of how specific differences exist in the culturally defined obstinacy of activities in these two fields of conduct, it can be expected that they change in a different manner in the wake of computer-mediated communication and globalisation. The increase in cultural reflexivity and relativisation of cultural characteristics through increased possibilities of comparison is of major importance here, too. In addition, globalisation processes increase the degree of networking of the activities included in globalisation processes, the culturally defined obstinacy of which changes through this to a greater or lesser extent.

The research field can be outlined by the following aspects:

- consequences of the convergence of electronic data processing technologies (computer) and computer-mediated communication (internet) for the societal (socio-economic, cultural) connections, forms and procedures of the handling of information and knowledge;
- relationship of the processes of information processing, storage and transmission and those of knowledge creation and provision, especially regarding the transfer of knowledge in science, economy and society as well as its use and exploitation (private – commercial);
- determination of knowledge in this context (definitions and conceptions of nature and functions; relationship between information and knowledge; knowledge as potential for action, differentiation of forms: cultural knowledge – scientific knowledge – professional knowledge – user knowledge; orientational knowledge versus instrumental knowledge, etc.); characteristics of changed forms of knowledge or a general change of form (acceleration of increase in knowledge and its devaluation; contextualisation);
- shifts in the relationship of the production factors knowledge, work, capital and nature – what makes knowledge the non-material key factor for the development of a European knowledge-based society;
- knowledge as a good versus knowledge as a common property (rights of ownership and use; copyright, patents);

- relationships between cultural knowledge and cultural diversity; consequences, problems and perspectives of the digitalisation in the context of preserving, passing on and accessing cultural heritage (cultural memory, cultural heritage) (education, libraries, museums, conservation of monuments).

5.2.4 Cross-sectional topic: Security/insecurity and trust

The research problem of the cross-sectional topic "security/insecurity and trust", which is at the same time the fourth research field of CULTMEDIA, concerns the changed balance of these two fundamental parameters for the constituency of modern societies, particularly since this problem unfolds in a special way in the intersectional plane of the three research fields mentioned above.

Every use of technology presupposes trust, for example regarding fulfilment of function, realisation of goals and availability. Here, trust can refer to individual people or more comprehensive social institutions and systems. While trust compensates the insecurities connected with every step, possibilities of conduct are opened up which would not have existed without this trust. In the German language, the word *Sicherheit*, which can be translated as *security*, is used in at least three senses (cf. Kaufmann 1970): security as a feeling of safety, as self-confidence and as system security (of means for various purposes which can be manufactured and calculated). All three uses are relevant for CULTMEDIA, since they refer to those human creations which are called new media. Thus the cultural way in which a society sees itself must be made the central topic, including its "understanding of security" (need for security, experience of insecurity, safety provisions, communication on possible advantages and disadvantages or "profits" and "losses"). In this process, the limit is fixed regarding the time and context-dependent acceptable and accepted conduct which uses technology (e.g., regarding the expenses for securing, the relationship of cost and benefit or the simplicity of handling), the stepping over of which may lead to individual or institutional "defence reactions" (refusal, inefficient use, reverting to conventional solutions, etc.). Beyond the individual need and desire for security, various social groups have developed different ways of handling insecurities, dangers and risks caused by technology. A solution to the problems related to individual and subjective needs for security can only lie in the development of appropriate security cultures in the irresolvable field of tension between individual and society.

This research field can be outlined by the following aspects:

- balance between the technical, legal, etc. measures for the protection of privacy and those for the protection of the public sphere;

- the problem of data protection and the measure of real risk; possible abuse of personal data and possible manipulation regarding the identity of people;
- internet culture and requirements for data protection and privacy;
- culture and insecurity – vulnerability of the modern-day infrastructure, with the internet serving as an example;
- consideration of the handling of questions of e-security within the area of the new media – security "cultures" in the narrow sense – by comparing countries and cultures;
- internet culture and requirements for data protection and privacy, the clear identification of the borders between the right of the individual to information and the right to privacy; censorship in the internet versus free access to complete information;
- culture and insecurity – vulnerability of modern-day infrastructures with the internet serving as an example; consideration of the handling of e-security questions within the surroundings of the new media – security "cultures" in the narrow sense – by comparing countries and cultures.

6 Synthetic-conciliar research level: Elements of European integration

By means of comparative investigation of the interactions of cultural diversity and new media in selected research fields, not only the patterns of the respective interactions are to be identified, but it should also be investigated to what extent this pattern varies when comparing European nations. Instead of "only" leading a general discussion on cultural diversity as an element of European integration, debating, e.g., if diversity should be understood as a strength or a weakness, the question of the connection between cultural diversity and new media is investigated in a particular way in the context of four concrete complexes, which are of special importance for the development of the present European society/societies and the progression of the EU integration processes.

The central matter of concern of CultMedia's "third" synthetic-conciliar research level is consequently to investigate the role of these interactions as an "element of European integration". In this context, first the importance of "cultural identity" (national identities – European identity) for the process of European integration has to be clarified, particularly in connection to its (respective) political and economic contextual conditions (cf. 6.1). Second, the task is to draw conclusions from the comparison of the interaction patterns between cultural diversity and new media mentioned above from the selected research fields (cf. 6.2). Finally, against this background, recommendations are to be developed for the further formation of the media and cultural policies of the EU and its member states. The problem in this context is not the coherence of its guidelines, but also the starting points and instruments as well as the question of whether, within the framework of the latest European expansion, there is a special need for action and what measures can meet this (cf. 6.3).

6.1 Considerations

Any discussion of questions regarding the connection of cultural diversity, new media and European integration presupposes an examination of the importance of "Europe".[12] Is this the name of a geographical place? A vision of the perfect civilisation? Or a cluster of interacting states which influence each other? At first glance it seems simple to decide in favour of none of the alternatives: Europe is a concept which is open to interpretation. In European history, "Europe" has adopted a number of different meanings and various preferences have been articulated. Especially for the present, which is in many ways characterised by radical changes and new departures throughout the world, it has to be assumed that "Europe" represents a changing concept.

An expression of the changing concept of "Europe" is the formulation of far-reaching and ambitious European-political development objectives, as, e.g., in the declaration of the Lisbon European Council of March 2000: "The Union has today set itself a new strategic goal for the next decade: to become the most competitive and dynamic knowledge-based economy in the world, capable of sustainable economic growth with more and better jobs and greater social cohesion" (European Council 2000).

Economic growth, employment, social cohesion, quality of life, and sustainable development can all be viewed as crucial aspects of a 21st century European model the EU is striving for. Taking into account Europe's diversity in this context is itself a feature of a specific European way of socio-economic development.

In the context of this European development model – that is also oriented towards similar ideas from the OECD (cf. Hodgson 2000) – the technology-

---

12 Here CULTMEDIA is trying to link up with the debates and results of two research projects, which have been funded and carried out within the framework of the European Science Foundation (ESF): "European identity – European diversity" (an ESF workshop project) and "Changing Media – Changing Europe" (an ESF scientific network) (cf. http://www.esf.org).

driven transformation process to a "knowledge-based society" is seen as the pacemaker. However, with the stress on "social cohesion" and "sustainability" a specifically European point is emphasised. Europe's cultural diversity plays an ambivalent role in this context. On the one hand it is considered as a problematic condition and obstacle for the further economic development of the entire common area. On the other hand it is considered to be a special strength because it bears considerable innovation potentials and provides, with the non-material economic factors "culture" and "knowledge", those resources whose value increases in a "knowledge-based society" in comparison with the other factors of productivity (capital, work, nature; cf. Stehr 1994). In this context CULTMEDIA sees it as its task to explicitly make the problem area of "cultural identity", "cultural diversity" and "European integration" its central topic, and this in connection with those new information and communication technologies that are considered the pacemakers of the "knowledge-based society".

The importance of this task is also emphasised by one of the most recognized newsletters of the European Science Foundation (ESF), which discusses the scientific-political and programmatic objectives of the Sixth Framework Plan of the EU commission; it says "Perhaps the unique and most self-evident contribution which the human sciences in Europe can make at this time concerns cultural identity, diversity and integration: namely for a sustainable European Union" (ESF 2002: 4). Furthermore, it states that "Within the context of the multicultural European Union and the future plans to fully integrate countries applying for accession, knowledge about and understanding of cultural identity, diversity and integration is vital to a successful integration process; for cultural change occurs on a different timescale, and often out of step with more rapid social and economic change" (ESF 2002: 4).

6.2 Conclusions

The following section provides some conclusions regarding the question of the role that the interaction patterns of cultural diversity and new media – especially in the areas of privacy and the public sphere, identity and the community, knowledge and the economy as well as security/insecurity and trust – play as elements of European integration.

The starting point is that the interactions of cultural diversity and new media as "elements" of European integration mentioned might play a contradictory role. Their ambivalence lies in the fact that they are suited to drive integration processes, but they are also connected with challenges which – if not mastered – can block integration processes or even cause disintegration processes.

In order to clarify these connections and to elaborate on results, CULTMEDIA discuss the following main questions:

1. Are these interaction patterns (for each of the topics 1, 2, 3 and the cross sectional topic 4) in the different European countries (in particular those of the participating research partners) similar or dissimilar to each other?
2. Can these (dis)similarities be distributed according to a geographic cultural area east/west or north/south pattern?
3. Do signs of a progressing "homogenisation" of European cultures result from the comparison of the interaction patterns and their possible distribution or do they indicate an increasing cultural "diversification"?
4. What can be deduced from the interaction patterns identified in the three plus one areas for the role of the new media in the process of European integration (under particular considering the latest European expansion)?

6.3 Recommendations

If the task is now to develop recommendations for the further shaping of the European media and cultural policies, this is done under the restriction that only the guidelines themselves constitute the topic and not the elaboration of explicit suggestions for the implementation of concrete single measures. Moreover, it is intended to illustrate the consequences regarding the guidelines from the research results and the conclusions drawn, especially regarding the shaping of the new information and communication media, particularly the internet. If in this context "cultural diversity" is used in the sense of a state worth preserving, this is done from a higher European perspective. The variety of different ethnic-national and language-related cultures is to be preserved. But what does this mean for the respective ethnic-national and language-related culture that is addressed as a unit in this variety? What does diversity mean from the respective national perspective? Does it mean conserving the national culture? Promoting its further development through intensified intercultural communication and the expansion of intercultural dialogues, even at the expense of its levelling during the mutual penetration of the European cultures?

In order to carry out a new, extended intercultural dialogue, new, extended communicative, media-related and cultural competences are needed. A central requirement is to learn how to cope appropriately with the growing extent and the increasing speed of present processes of change and their consequences.

In mixed-national internet "communities" – the hypothesis can be formulated in this way – there are good prerequisites for the expansion of intercultural communication in the sense of intercultural dialogue. In such transcultural spaces there are opportunities to further develop communicative, media-related and (inter)cultural competences simultaneously or combined with each other.

In order not to lose the chances of the information society both the homogenisation and differentiation tendencies are to be considered adequately and used as opportunities for shaping the European future.

7 Outlook: Pursuits and wishes

The complexity and dynamics of the "quadratic" relationship of the intensive and far-reaching interactions of culture, society, technology, and the media is not only of major importance for Europe's future. It has also resulted in a research field that goes "beyond" the disciplinary specialisations of individual experts. In addition to this, developing such a project also needs intercultural cooperation.

The European research-network CultMedia, whose design has been presented here, embraces at present 16 partner-institutes out of 11 different European countries. During the ten years of the network's existence, an increasing convergence is recognizable among the methods and theories that determine the research-practice of all participants. Moreover, one can perceive an increasing interdisciplinary consciousness as against the leading questions and focal problems of the network. In order to strengthen the network and its research capacities, more partners are needed in order to incorporate them into the consortium. New elements of the net are desired and necessary in order to enlarge the theoretical as well as empirical basis of the investigations. Interested research-groups and competent (single) persons are friendly invited to take part and contact us.

In the context of CultMedia and the activities of its partners a series of meetings and conferences have been organized and realised so far, among others in Berlin (D), Prague (CZ), Potsdam (D), Banská Bystrica (SR), Katowice (PL), Budapest (H), San Sebastián (E) and Klagenfurt (A). The proceedings of these reunions have been published already or are about to be published.[13] Currently, the net-

13 Published results (as a selection) are the following: Banse 2003, 2005; Banse and Bartiková 2004, 2006; Banse and Kiepas 2008; Banse and Krebs 2011; Banse, Winter and Wieser 2009; Fleissner and Romano 2006; Galántai, Petsche and Várkonyi 2006; Hauser

work is occupied with several activities focusing on the realization and specification of its investigational tasks in the following areas of research:

(a) in the context of CULTMEDIA's research on "knowledge and the economy": "media-education and media-competence" (with a leading role of the network-node Katowice) and "e-administration" (with a leading role of the network-node Cottbus);

(b) in the context of CULTMEDIA's research on "(in-)security and trust": „security-cultures" (with a leading role of the network-nodes Karlsruhe and Wuppertal; cf. e.g., Banse and Hauser 2008a, 2008b)

(c) in the framework of the ongoing debate about the common grounds of all of CULTMEDIA's research fields:
"theoretical foundations", above all concerning the description of the general relations of culture and technology as well as the general characterisation of the information-society (with a leading role of the network-nodes Karlsruhe, Münster and Potsdam; cf. e.g., Metzner-Szigeth 2005, 2006, 2007, 2008).

Currently the European research network CULTMEDIA undergoes a process of consolidation. CULTMEDIA is engaged in the ignition of bi- and multilateral research-projects, in the exchange of scientific stuff, and in the promotion of young scientists.

References

Appadurai, Arjun

1996 *Modernity at Large. Cultural Dimensions of Globalization* (4th ed.). Minneapolis: University of Minnesota Press.

Banse, Gerhard

1999 Die Verbindung "wahrer Grundsätze" und "zuverlässiger Erfahrungen". Zur Möglichkeit und Wirklichkeit von Allgemeiner Technikwissenschaft nach Johann Beckmann. In: Bayerl, Günter and Beckmann, Jürgen (eds.), *Johann Beckmann (1739–1811). Beiträge zu Leben, Werk und Wirkung des Begründers der Allgemeinen Technologie*, 329–350. Münster: Waxmann.

Banse, Gerhard (ed.)

2003 "Cultural Diversity and New Media." Thematisches Heft der Zeitschrift Teorie Vědy. Časopis pro teorii vědy, techniky a komunikace/Theory of Science. *Journal for Theory of Science, Technology & Communication*, vol. 24, no. 1.

2010; Kiepas, Sułkowska and Wołek 2009; Kiepas, Szczepański and Żydek-Bednarczuk 2006; Kiepas and Żydek-Bednarczuk 2006; Petsche 2005; Petsche and Zapf 2008; Petsche, Zapf and Köhler 2007; Ursua and Metzner-Szigeth 2006; Ursua, Metzner-Szigeth and Gonzáles 2012.

Banse, Gerhard (ed.)
2005 *Neue Kultur(en) durch Neue Medien (?). Das Beispiel Internet.* Berlin: trafo.

Banse, Gerhard and Bartíková, Monika (eds.)
2004 "E-Learning Issues." Thematisches Heft der Zeitschrift Teorie Vědy. Časopis pro teorii vědy, techniky a komunikace/Theory of Science. *Journal for Theory of Science, Technology & Communication*, vol. 25, no. 3.

Banse, Gerhard and Bartíková, Monika (eds.)
2006 *e-Learning? – e-Learning!* Berlin: trafo.

Banse, Gerhard, Grunwald, Armin and Rader, Michael (eds.)
2002 *Innovations for an e-Society. Challenges in Technology Assessment.* Berlin: edition sigma.

Banse, Gerhard and Hauser, Robert
2008a Technik und Kultur. Das Beispiel Sicherheit und Sicherheitskultur(en). In: Rösch, Olga (ed.), *Technik und Kultur*, 61–83. Berlin: News & Media/Marcus v. Amberg.

Banse, Gerhard and Hauser, Robert
2008b Technik als (Intra- und Inter-)Kulturelles. Exemplarisches. In: Gronau, Norbert and Eversheim, Walter (eds.), *Umgang mit Wissen im interkulturellen Vergleich. Beiträge aus Forschung und Unternehmenspraxis*, 49–77. Stuttgart: Fraunhofer IRB.

Banse, Gerhard and Kiepas, Andrzej (eds.)
2008 *Visionen der Informationsgesellschaft 2016.* Berlin: trafo.

Banse, Gerhard and Krebs, Irene (eds.)
2011 *Kulturelle Diversität und Neue Medien. Entwicklungen – Interdependenzen – Resonanzen.* Berlin: trafo.

Banse, Gerhard, Langenbach, Christian J. and Machleidt, Petr (eds.)
2000 *Towards the Information Society. The Case of Eastern European Countries.* Berlin/Heidelberg: Springer.

Banse, Gerhard, Winter, Rainer and Wieser, Matthias (eds.)
2009 *Neue Medien und kulturelle Vielfalt. Konzepte und Praktiken.* Berlin: trafo.

Beck, Ulrich
1997 *Was ist Globalisierung?* Frankfurt am Main: Suhrkamp.

Boehnke, Klaus, Dilger, Werner, Habscheid, Stephan, Holly, Werner, Keitel, Evelyne, Krems, Josef, Münch, Thomas, Schmied, Josef, Stegu, Martin and Voß, Günter
1999 *Neue Medien im Alltag: Von individueller Nutzung zu soziokulturellem Wandel.* Lengerich: Pabst Science Publishers.

Döring, Nicola
1998 *Sozialpsychologie des Internet.* Göttingen: Hogrefe.

Council, European
2000 *Presidency Conclusions, Lisbon Summit*, 23–24. March2000. – URL: http://www.europarl.eu.int/summits/lis1_en.htm [03. 03. 2004].

European Science Foundation (ESF)
2002 Reflections. (*Newsletter of the Standing Committee for the Humanities*, 6.) Strasbourg.

Faulstich, Werner
2000 Medium. In: Faulstich, Werner (ed.), *Grundwissen Medien* (4th ed.), 21–108. München: Wilhelm Fink.

Fleissner, Peter and Romano, Vicente (eds.)
2006 *Digitale Medien – Neue Möglichkeiten für Demokratie und Partizipation?* Berlin: trafo.

Fuchs, Max
1999 *Mensch und Kultur. Zu den anthropologischen Grundlagen von Kulturarbeit und Kulturpolitik*. Opladen: Westdeutscher Verlag.
Galántai, Zoltán, Petsche, Hans-Joachim and Várkonyi, László (eds.)
2006 *Internet Security. Facetten eines Problems*. Berlin: trafo.
Gehlen, Arnold
1958 *Der Mensch. Seine Natur und seine Stellung in der Welt* (6th ed.). Bonn: Athenäum.
Giddens, Anthony
1990 *The Consequences of Modernity*. Stanford, CA.: Stanford University Press.
Grunwald, Armin, Gerhard Banse, Christopher Coenen and Hennen, Leomhard
2006 *Netzöffentlichkeit und digitale Demokratie. Tendenzen politischer Kommunikation im Internet*. Berlin: edition sigma.
Habermas, Jürgen
1981a *Theorie des kommunikativen Handelns. Bd. 1: Handlungsrationalität und gesellschaftliche Rationalisierung*. Frankfurt am Main: Suhrkamp.
Habermas, Jürgen
1981b *Theorie des kommunikativen Handelns. Bd. 2: Zur Kritik der funktionalistischen Vernunft*. Frankfurt am Main: Suhrkamp.
Habermas, Jürgen
1989 *The Structural Transformation of the Public Sphere*. Cambridge, MA/London: Massachusetts Institute of Technology.
Hauser, Robert
2010 *Technische Kulturen oder kultivierte Technik. Das Internet in Deutschland und Russland*. Berlin: trafo.
Hermeking, Marc
2001 *Kulturen und Technik. Techniktransfer als Arbeitsfeld der Interkulturellen Kommunikation. Beispiele aus der arabischen, russischen und lateinamerikanischen Region*. Münster: Waxmann.
Hodgson, Geoffrey M.
2000 Socio-economic Consequences of the Advance of Complexity and Knowledge. In: OECD (ed.), *The Creative Society of the 21st Century*, 89–112. Paris: OECD.
Hörning, Karl. H.
1985 Technik und Symbol. Ein Beitrag zur Soziologie alltäglichen Technikumgangs. *Soziale Welt*, vol. 36, no. 2, 185–207.
Kaufmann, Franz-Xaver
1970 *Sicherheit als soziologisches und sozialpolitisches Problem. Untersuchungen zu einer Wertidee hochdifferenzierter Gesellschaften*. Stuttgart: Enke.
Kiepas, Andrzej, Sułkowska, Mariola and Wołek, Magdalena (eds.)
2009 *Człowiek a światy wirtualne [Human and virtual Reality]*. Katowice: Verlag der Schlesischen Universität.
Kiepas, Andrzej, Szczepański, Marek S. and Żydek-Bednarczuk, Urszula (eds.)
2006 *Internet – Społeczeństwo informacyjne – Kultura [Internet – Global Communication – Identity]*. Tychy: Verlag der Hochschule für Wirtschaft und Sozialwissenschaften.
Kiepas, Andrzej and Żydek-Bednarczuk, Urszula (eds.)
2006 *Informationsgesellschaft und Kultur. Internet – Globale Kommunikation – Identität*. Berlin: trafo.

King, Anthony D.
1993 Introduction. Spaces of Cultures, Spaces of Knowledge. In: King, Anthony D. (ed.), *Culture, Globalization and the World-System. Contemporary Conditions for the Representation of Identity,* 1–18, Binghamton, NY.: State University of New York.
Klein, Hans-Joachim
2000 Kultur. In: Schäfers, Bernhard (eds.), *Grundbegriffe der Soziologie* (6th ed.), 196–199. Opladen: Leske + Budrich.
Kroeber, Alfred L. and Kluckhohn, Clyde
1952 *Culture. A Critical Review of Concepts and Definitions.* Cambridge, MA: Vintage Books.
Langenbach, Christian J. and Ulrich, Otto (eds.)
2002 *Elektronische Signaturen. Kulturelle Rahmenbedingungen einer technischen Entwicklung.* Berlin/Heidelberg: Springer.
McIntosh, Donald
1997 Language, Self, and Lifeworld in Habermas's Theory of Communicative Action. In: Boudon, Reymond, Cherkaoui, Mohammed and Alexander, Jeffrey C. (eds.), *The Classical Tradition in Sociology. The European Tradition. Vol. IV: European Sociology in the Twentieth Century: From Post-Industrialism to Post-Modernism,* 102–130, London/Thousand Oaks/New Delhi: Sage.
Metzner-Szigeth, Andreas
2005 Knowledge, Computer-Mediated Communication and Sustainable Development. In: Banse, Gerhard, Hronszky, Imre and Nelson, Gordon (eds.), *Rationality in an Uncertain World,* 251–261. Berlin: edition sigma.
Metzner-Szigeth, Andreas
2006 "El movimiento y la matriz" – Internet y transformación socio-cultural. *Revista Iberoamericana de Ciencia, Tecnología, Sociedad e Innovación (CTS+I),* no. 7, – URL: http://www.campus-oei.org/revistactsi/numero7/articulo04.htm.
Metzner-Szigeth, Andreas
2007 Internet & Gesellschaft: Ein Humanes Projekt? *Sic et Non – Zeitschrift für Philosophie und Kultur – im Netz,* no. 8, – URL: http://www.sicetnon.org/content/pdf/internet&gesellschaft.pdf.
Metzner-Szigeth, Andreas
2008 Von Cyber-Identitäten, virtuellen Gemeinschaften und vernetzter Individualisierung – sozial-psychologische Überlegungen. *Sic et Non – Zeitschrift für Philosophie und Kultur – im Netz,* no. 9, – URL: http://sicetnon.org/content/pdf/cyber-ident.pdf.
Paschen, Herbert, Wingert, Bernd, Coenen, Christopher and Banse, Gerhard
2002 *Kultur – Medien – Märkte. Medienentwicklung und kultureller Wandel.* Berlin: edition sigma.
Petsche, Hans-Joachim (ed.)
2005 *Kultur und/oder/als Technik – zur frag-würdigen Medialität des Internets.* Berlin: trafo.
Petsche, Hans-Joachim and Zapf, Antje
2008 *Die Neuen Medien und die kulturelle Vielfalt Europas. Eine empirisch vergleichende Erhebung unter Studierenden Deutschlands, Polens, Spaniens, Tschechiens und Ungarns (unter besonderer Berücksichtigung des Internets).* Berlin: trafo.

Petsche, Hans-Joachim, Zapf, Antje and Köhler, Thomas
2007 *Die Neuen Medien und die kulturelle Vielfalt Europas. Empirisch-vergleichende Erhebung unter Studierenden Deutschlands, Polens, Spaniens, Tschechiens und Ungarns*. 2 vols. Berlin: trafo.

Rammert, Werner
2000 Virtuelle Realitäten als medial erzeugte Sonderwirklichkeiten. Veränderungen der Kommunikation im Netz der Computer. In: Rammer, Werner (ed.), *Technik aus soziologischer Perspektive II*, 115–128. Opladen: Westdeutscher Verlag.

Robertson, Roland
1998 Glokalisierung: Homogenität und Heterogenität in Raum und Zeit. In: Beck, Ulrich (ed.), *Perspektiven der Weltgesellschaft*, 192–220. Frankfurt am Main: Suhrkamp.

Ropohl, Günter
1993 Technik. In: *Brockhaus-Enzyklopädie*, Vol. 21, 672–674. Mannheim: Brockhaus.

Sachsse, Hanns
1992 Technik. In: Seiffert, Helmut and Radnitzky, Gerard (eds.), *Handlexikon zur Wissenschaftstheorie*, 358–361. München: Ehrenwirth.

Schmidt, Siegfried J.
2000 *Kalte Faszination. Medien, Kultur, Wissenschaft in der Mediengesellschaft.* Weilerswist: Velbrück.

Slevin, James
2000 *The Internet and Society*. Cambridge: Wiley-Blackwell.

Stehr, Nico
1994 *Knowledge Societies*. London: Sage.

trafo Verlagsgruppe Dr. Wolfgang Weist. – URL: http://www.trafoberlin.de.

UNESCO
1982 Mexico City Declaration on Cultural Policies. World Conference on Cultural Policies. Mexico City, 26 July – 6 August 1982 – URL: http://portal.unesco.org/culture/en/files/12762/11295421661mexico_en.pdf/mexico_en.pdf.

Ursua, Nicanor and Metzner-Szigeth, Andreas (eds.)
2006 *Netzbasierte Kommunikation, Identität und Gemeinschaft/Net Based Communication, Identity and Community*. Berlin: trafo.

Ursua, Nicanor, Metzner-Szigeth, Andreas and de Dios Gonzáles, Juan (eds.)
2012 *e-Epistemología, Metaconocimiento y Virtualidad*. México [in print].

Biographical notes

Gerhard Banse (born in 1946), studied chemistry, biology, and pedagogic at Teacher Training College Potsdam (Germany), and philosophy at Humboldt University Berlin (Germany); 1969 diploma; 1974 dissertation (Dr. phil.) and 1981 habilitation (Dr. sc. phil.) on philosophical problems of technology; 1988 professor for philosophy; since 1974 scientist and senior at Institute of Philosophy of Academy of Sciences of GDR, Department of Philosophy of Technology at Technical University Cottbus (Germany), European Academy Bad Neuenahr-Ahrweiler (Germany), Institute for Philosophy at University of Potsdam and Institute for Technology Assessment and Systems Analysis at Karlsruhe Institute of Technology (KIT; Germany); professor e.h. at Silesian University of Katowice (Poland), guest professor at Matej Bel-University Banská Bystrica (Slovak Republic) and professor of general technology at Technical University of Brandenburg Cottbus (Germany); chairman of Leibniz Sozietät of Sciences and Humanities Berlin (Germany). Research fields are philosophy of technology and general technology, lectures in the domains history and philosophy of technology, general risk research, technological change, theory of technical sciences and technology assessment. Contact: gerhard.banse@kit.edu

Colette Cortès started her academic carrier at the Université Paris Sorbonne (Paris IV, 1970–1989, from 1977 as Maître de conférences) and was a professor of Applied Linguistics and German as Foreign Language at the University Paris Diderot (Paris VII; from 1989–2007); she was Head of the Institution EILA: Études Interculturelles de Linguistique Appliquée (1993–2002) and Director of the Research Centre CIÉL: Centre Interlangue d'Études en Lexicologie (1992–2006). From 1998–2001 she was Vice-Rector of Paris VII. Her research interest is in lexicology and terminology of French and German, in general and specialised text linguistics as well as in syntax, semantics and pragmatics. 1970 she received her PhD on studies concerning the subject in German and in 1988 she did her habilitation (Syntax and Pragmatics of Subordination in German). She is author of a variety of papers and books and co-author of a dictionary of French Nominal-Predicates (2009, together with Thomas Kotschi and Ulrich Detges) and actually editor of a book on Sentence Opening. Form, Function, Strategy, 2012, Tübingen. Contact: colette.cortes@free.fr

Franck Ganier received his PhD in Cognitive Psychology in 1999 at the University of Burgondy, Dijon, France. He is now a professor of Cognitive Psychology and Ergonomics at the European University of Brittany, University of Brest, France. He is a member of the Lab-STICC Laboratory (UMR 6285 CNRS), European Centre for Virtual Reality (CERV), Brest, France. His research interests concern procedural learning, processes for comprehending instructions and methods for evaluating technical documents. Franck Ganier is leader of the working group "Design of written documents in the workplace" of the COST Action "European Research Network on Learning to Write Effectively". Contact: franck.ganier@univ-brest.fr

Marc Hermeking is scientific lecturer at the Institute for Intercultural Communications at the Ludwig-Maximilians University (LMU) Munich, Germany. He studied business administration (diploma) and marketing psychology, European ethnology and intercultural communications (doctor). His doctoral thesis deals with cultural influences on the international transfer and usage of technology (Kulturen und Technik, published 2001). His teachings focus on Intercultural Marketing with special interest in cross-cultural product development and marketing communications, including human-machine interaction and Website design, as well as the transfer of technology into foreign cultures. Additionally, he works as a trainer and lecturer for diverse industrial corporations and educational institutions. He has published several contributions for textbooks and journals. Contact: Marc.Hermeking@t-online.de

Andreas Metzner-Szigeth Education: Studies in the Social Sciences, Biology and Philosophy; Dr. phil. (1991); Habilitation in Sociology (2000) Current position: Adjunct Professor (since 2001) at the Institute of Sociology of the Univ. of Münster, Germany, as well as Executive Manager and Consultant (since 2008) at the Episteme Consulting, Münster, Germany; Visiting Professor at the Univ. of the Basque Country in San Sebastián, Spain (since 2004), the Univ. of Katowice, Poland (2009–12), and the Univ. of Vienna, Austria (in 2007) Areas of Research and Teaching: Studies of Science, Technology & Society (STS); Systems-Analysis & Technology Assessment; Risk-Analysis & Vulnerability-Studies; Environmental Sociology & Sustainability Science; Investigations in Media, Communication & Culture; Political Sociology & Governance Research; Economic Sociology & Management Studies. Web: http://egora.uni-muenster.de/ifs/metzner.shtml – Contact: metzner@uni-muenster.de

Karl-Heinz Pogner, holds a Ph.d. from Odense University (Denmark) on the basis of his thesis "Schreiben im Beruf als Handeln im Fach" [Writing in the workplace as acting in the professions] (1999). He has lectured at Odense University (Denmark) and the Southern Denmark Business School in Sønderborg (Denmark). In 1998 he joined the Department of Intercultural Communication and Management, Copenhagen Business School, as Associate Professors. He is the Academic Program Director of the B.SC. and M.Sc. study programs "Business Administration and Organizational Communication" and one of the founding persons of the "ComCaseCompetion". His main research and teaching interests are in the areas of text production, writing in the workplace, corporate communication, and organizational communication. Contact: kp.ikl@cbs.dk

Annely Rothkegel is a guest professor at the University of Hildesheim, formerly professor of Applied Linguistics (and Technical Communication) at the University of Technology, Chemnitz (2003–2009) and at the University of Applied Sciences of Hannover (1994–2003); PhD in linguistics in 1973, and Habilitation in General Linguistics and Computational Linguistics at the University of Saarland (Saarbrücken, Germany) in 1992. Topics of research cover semantic models, knowledge and text work, hypertext, machine translation, discourse analysis and text production. Has participated in a number of large-scale, national and international research projects on these topics, and developed curricula for graduate and post-graduate studies in technical writing, e.g. EU-LEONARDO-program, the Finnish project COLC (Computing for Language Carriers), BA Technical Communication (University of Technology, Chemnitz). Co-Editor of the series TTCP (Mouton de Gruyter). Contact: annely.rothkegel@uni-hildesheim.de

Sonja Ruda works at the Chair of Applied Linguistics/Communication on Technology at Chemnitz University of Technology, Germany. She studied German Language and Literature and Computer Linguistics at Trier University. Her doctoral dissertation "Teacher assisting feedback tools for online courses" was part of the research project PortaLingua, an internet portal for learning and teaching linguistics and communication science online. Her further research interests include pragmalinguistics, computational linguistics, concepts of questions and e-learning. Contact: sonja.ruda@phil.tu-chemnitz.de

Vasco Alexander Schmidt, Dr., is a mathematician, linguist, and technical writer working for SAP Research, the global technology research and innovation unit of SAP. SAP is the market leader in enterprise application software. Vasco Schmidt is member of the interdisciplinary research network "Language and Knowledge – problems in public and professional communication", where he leads the domain mathematics. He is a trainer for creative and technical writing at SAP and has been lecturer at University of Heidelberg and University of Siegen (Germany). His research interests include software documentation, technical writing, and innovation communication. Contact: schmidt.vasco@web.de

Klaus Schubert is a Professor of Applied Linguistics/International Specialized Communication at the University of Hildesheim (Germany). He was a sociolinguist at the University of Kiel (Germany), a computational linguist and project leader at the Research Department of the software house BSO/Buro voor Systeemontwikkeling BV in Utrecht (Netherlands) and a consultant and technical manager at BSO/Language Technology BV in Baarn (Netherlands). For many years he served as a Professor of Computational Linguistics and Technical Communication at Flensburg University of Applied Sciences in Flensburg (Germany). Main research areas: Applied Linguistics, Translation Studies, Specialized Communication Studies, Interlinguistics, Applied Computer Linguistics. Recent book: *Wissen, Sprache, Medium, Arbeit* (Tübingen: Narr 2007). Contact: klaus.schubert@uni-hildesheim.de

Clemens Schwender studied German literature, philosophy, psychology and media science at TU Berlin, receiving his doctorate in Philosophy (about the history of technical documentation in the Berlin telephone directory) there in 1996, and his Habilitation in Media Science (about media and emotions) in 2001. Prior to joining University for Popular Arts (FH) in 2011 as Professor of Media Psychology he was Professor of Communication Science at Jacobs University Bremen, as well as at University of Management and Communication and at Business School Potsdam. His research topics include media and emotions, lifelong learning, audio-visual argument, advertisement, and media history. Contact: clemens.schwender@googlemail.com

Index

Bei Fragen zur Produktsicherheit wenden Sie sich bitte an:
If you have any questions regarding product safety,
please contact:

Walter de Gruyter GmbH
Genthiner Straße 13
10785 Berlin
productsafety@degruyterbrill.com